NATIONAL 5

PHYSICS
WITH ANSWERS

SECOND EDITION

Arthur Baillie

HODDER
GIBSON
AN HACHETTE UK COMPANY

The Publishers would like to thank the following for permission to reproduce copyright material:

Photo credits

p.1 (background) and Section 1 running head image © Zhu difeng/stock.adobe.com (inset C) © Arthur Baillie (inset R) © EPSTOCK/stock.adobe.com; **p.3** (top) © GIPHOTOSTOCK/SCIENCE PHOTO LIBRARY (bottom) © Mike Goldwater / Alamy Stock Photo; **p.5** © Arthur Baillie; **p.7** © Arthur Baillie; **p.9** (top) © TREVOR CLIFFORD PHOTOGRAPHY/SCIENCE PHOTO LIBRARY (bottom) © EPSTOCK/stock.adobe.com; **p.19** © MARTYN F. CHILLMAID/ SCIENCE PHOTO LIBRARY; **p.26** (top) © Arthur Baillie (bottom) © Arthur Baillie; **p.27** © Arthur Baillie; **p.37** (background) and Section 2 running head image © derek82/stock.adobe.com (inset C) © lazarovphoto/stock.adobe.com (inset R) © Arthur Baillie; **p.40** (top left) © NASA/JPL-Caltech/R. Hurt (SSC) (bottom left) © NASA/JPL-Caltech/R. Hurt (SSC) (centre right) © DR RAY CLARK & MERVYN GOFF/SCIENCE PHOTO LIBRARY; **p.41** © JHP Attractions / Alamy Stock Photo; **p.50** (top) © lazarovphoto/stock.adobe.com (bottom) © Ruslan Kurbanov/stock.adobe.com; **p.51** © Arthur Baillie; **p.63** (background) and Section 3 running head image © Thomas/stock.adobe.com (inset L) © Ttstudio/stock.adobe.com (inset C) © Cn Boon / Alamy Stock Photo; **p.67** © Ttstudio/stock. adobe.com; **p.68** © Cn Boon / Alamy Stock Photo; **p.79** (background) and Section 4 running head image © sakkmesterke/stock.adobe.com (inset C) © MARTIN DOHRN/SCIENCE PHOTO LIBRARY; **p.88** (top left) © JOSH SHER/SCIENCE PHOTO LIBRARY (top right) © MARTIN DOHRN/SCIENCE PHOTO LIBRARY; **p.89** © MATT MEADOWS/SCIENCE PHOTO LIBRARY; **p.103** (background) and Section 5 running head image © Alex Koch/stock.adobe.com (inset L) © Anatoly Vartanov/stock.adobe.com (inset R) © Antony Nettle / Alamy Stock Photo; **p.105** © Arthur Baillie; **p.112** © Arthur Baillie; **p.117** © sciencephotos / Alamy Stock Photo; **p.120** © Stock Connection Blue / Alamy Stock Photo; **p.121** (top left) © SKYSCAN/SCIENCE PHOTO LIBRARY (bottom left) © Anthony Kay/Archive / Alamy Stock Photo (centre right) © Anatoly Vartanov/stock.adobe.com; **p.123** © Arthur Baillie; **p.134** (top right) © Antony Nettle / Alamy Stock Photo (bottom right) © Worldspec/NASA / Alamy Stock Photo; **p.147** (background) and Section 6 running head image © vj/stock.adobe.com (inset L) © Elena Schweitzer/ stock.adobe.com (inset C) © h3llb0y66/stock.adobe.com (inset R) © Mark Williamson/stock.adobe.com; **p.148** (left) © Elena Schweitzer/stock. adobe.com (right) © h3llb0y66/stock.adobe.com; **p.149** © stuart thomson / Alamy Stock Photo; **p.150** © NASA/SCIENCE PHOTO LIBRARY; **p.151** (left) © European Space Agency (right) © PLANETOBSERVER/SCIENCE PHOTO LIBRARY; **p.152** (top left) © NASA/SCIENCE PHOTO LIBRARY (bottom left) © Athol Pictures / Alamy Stock Photo; **p.159** (left) © Mark Williamson/stock.adobe.com (right) © Cosmo Condina North America / Alamy Stock Photo; **p.161** (top left) © NASA/ CXC/ESA/ASU/SCIENCE PHOTO LIBRARY (bottom left) © NRAO/AUI/NSF/SCIENCE PHOTO LIBRARY (top right) © NASA/ESA/STSCI/J.MORSE, U.COLORADO/ SCIENCE PHOTO LIBRARY; **p.162** (right) © Michael Lemke and Simon Jeffery; **p.163** © SHEILA TERRY/SCIENCE PHOTO LIBRARY.

Acknowledgements The *Useful Physics equations* and *Data Sheet* are reproduced with kind permission of SQA.

Every effort has been made to trace all copyright holders, but if any have been inadvertently overlooked, the Publishers will be pleased to make the necessary arrangements at the first opportunity.

Whilst every effort has been made to check the instructions of practical work in this book, it is still the duty and legal obligation of schools to carry out their own risk assessments.

Although every effort has been made to ensure that website addresses are correct at time of going to press, Hodder Gibson cannot be held responsible for the content of any website mentioned in this book. It is sometimes possible to find a relocated web page by typing in the address of the home page for a website in the URL window of your browser.

Hachette UK's policy is to use papers that are natural, renewable and recyclable products and made from wood grown in well-managed forests and other controlled sources. The logging and manufacturing processes are expected to conform to the environmental regulations of the country of origin.

Orders: please contact Hachette UK Distribution, Hely Hutchinson Centre, Milton Road, Didcot, Oxfordshire, OX11 7HH. Telephone: +44 (0)1235 827827. Email education@hachette.co.uk Lines are open from 9 a.m. to 5 p.m., Monday to Friday. You can also order through our website: www.hoddereducation.co.uk. If you have queries or questions that aren't about an order, you can contact us at hoddergibson@hodder.co.uk

First published in 2013 © Arthur Baillie

This second edition published in 2018 by
Hodder Gibson, an imprint of Hodder Education,
An Hachette UK Company
50 Frederick Street
Edinburgh EH2 1EX

Without Answers

Impression number	5	4	3	2	1
Year	2022	2021	2020	2019	2018

ISBN: 978 1 5104 2927 7

With Answers

Impression number	5	4	3
Year	2025	2024	2022

ISBN: 978 1 5104 2928 4

Cover photo: © Nik_Merkulov - stock.adobe.com

Typeset in Minion Regular 11/14 by Integra Software Services Pvt. Ltd., Pondicherry, India

Printed and bound by CPI Group (UK) Ltd, Croydon, CR0 4YY

A catalogue record for this title is available from the British Library

Contents

Preface

This book is designed to act as a valuable resource for students studying SQA National 5 Physics. It provides a core text that adheres closely to the SQA specification. Each section of the book matches a mandatory component of the specification. Each chapter corresponds to a number of mandatory course key areas.

In addition to the core text, the book contains a variety of special features: a list of the knowledge and understanding covered in each chapter, worked examples, physics beyond the classroom, key facts and physics equations, and end-of-chapter questions.

Three sets of examination-style questions are spread through the book to foster the development of the mandatory subject skills outlined in the course arrangements. The questions are designed to prepare students for National 5 assessment, where they will be expected to demonstrate their ability to solve problems, select relevant information, present information, process data, plan experimental procedure, evaluate experimental design, draw valid conclusions and make predictions and generalisations.

As detailed on the inner front cover of this book, always check **www.sqa.org.uk** for the most up-to-date specifications.

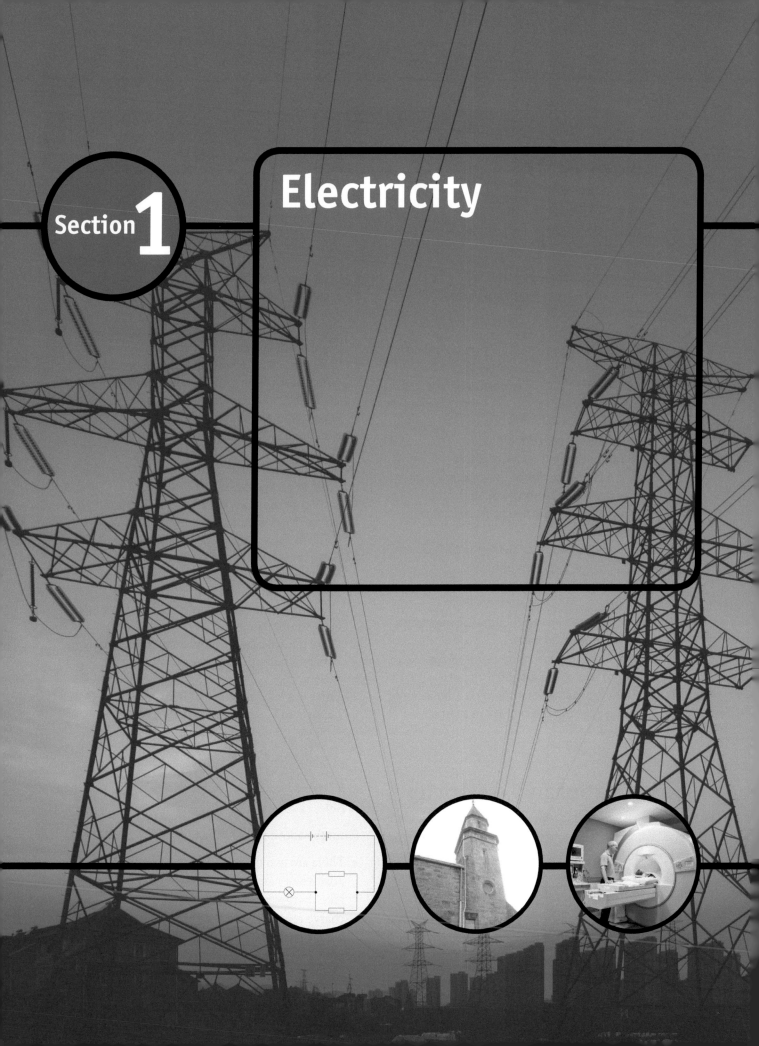

Section **1**

Electricity

1 Electrical circuits

Learning outcomes

At the end of this chapter you should be able to:

1 State that there are two types of charge: positive (+) and negative (−).
2 State that:
   a) charges that have the same sign are called 'like charges'
   b) charges that have the opposite sign are called 'unlike charges'.
3 State that:
   a) like charges repel each other
   b) unlike charges attract each other.
4 Describe a simple model of the atom that includes protons, neutrons and electrons.
5 State that in an electric field a charged object experiences a force.
6 Carry out calculations involving the relationship between charge, current and time.
7 State that electrons are free to move in a conductor.
8 Distinguish between conductors and insulators and give examples of each.
9 Describe electrical current in terms of the movement of charges around a circuit.
10 State that the voltage of a supply is a measure of the energy given to the charges in a circuit.
11 Draw and identify circuit symbols for a cell, battery, resistor, variable resistor, switch, lamp, ammeter and voltmeter.
12 Draw circuit diagrams to show the correct positions of an ammeter and voltmeter in a circuit.
13 State that V/I for a resistor remains constant for different currents provided the temperature of the resistor remains constant.
14 State that an increase in the resistance of a circuit leads to a decrease in the current in that circuit.
15 State that in a series circuit the current is the same at all positions.
16 State that the sum of the potential differences across the components connected in series is equal to the voltage of the supply.
17 State that the sum of the currents in parallel branches is equal to the current from the supply.
18 State that the potential difference across components connected in parallel is the same for each component.
19 Carry out calculations involving the relationship between potential difference, current and resistance.
20 Carry out calculations involving resistors connected in series and parallel.

Static (stationary) electricity

Some materials, such as acetate and polythene rods, when rubbed with a cloth are said to become charged. A charged rod is able to lift small pieces of paper (Figure 1.1) and can attract or repel another charged rod. Experimenting with charged rods leads to the following conclusions:

- A charged acetate rod repels a charged acetate rod.
- A charged acetate rod attracts a charged polythene rod.
- A charged polythene rod repels a charged polythene rod.

From the above we can conclude that:

- There are two types of charge. They are called positive (+) and negative (−).
- Like charges repel (charges with the same sign of charge repel).
- Unlike charges attract (charges with the opposite sign of charge attract).

Figure 1.1 A charged rod attracts small pieces of paper

Figure 1.2 Like charges on a person – a hair-raising experience!

All solids, liquids and gases are made up of atoms. An atom consists of a positively charged centre or nucleus surrounded by a 'cloud' of rapidly revolving negative charges called electrons. The nucleus consists of particles called protons (positively charged) and neutrons (uncharged) – see Figure 1.3.

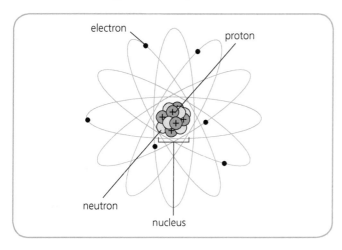

Figure 1.3 Model of an atom

Charge is measured in coulombs (C). The charge on a proton is 1.6×10^{-19} C. The charge on an electron has the same size of charge as a proton but it is negative i.e. -1.6×10^{-19} C.

When a rod is rubbed with a cloth, it is only the outer electrons of the atoms of the rod and cloth that are affected. With some materials, the outer electrons are held tightly by the nucleus while for other materials the outer electrons are held loosely by the nucleus. This means that when a rod is rubbed with a cloth, the atoms of the rod 'lose' electrons to the cloth or the rod 'gains' electrons from the atoms of the cloth; there is a transfer of electrons (negative charges) from one material to the other.

- When an object gains electrons it becomes negatively charged.
- When an object loses electrons it becomes positively charged.

An electric field

When a small positive charge is placed near a charged object, it is either attracted or repelled, and so it experiences a force. The region surrounding the charged object, where another charged object experiences a force, is called an **electric force field** or simply an **electric field**.

A charged object cannot exert a force on itself as it cannot experience its own electric field.

An electric field line is the line along which a small positive test charge would move if free to do so. Electric field lines represent the direction of the electrical force on a positive charge.

- The lines of force are continuous – starting on a positive charge and ending on a negative charge.
- The lines never touch or cross (like contours on a map).
- The closer together the lines are, the stronger the electric field.
- The arrows on the lines of force always point from positive to negative, i.e. they show the direction that a positive charge would move in if it were free to move.

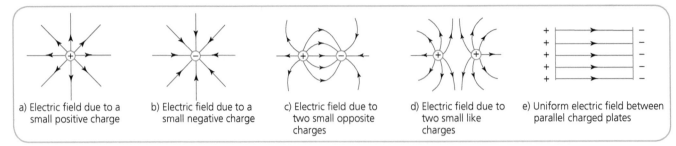

a) Electric field due to a small positive charge

b) Electric field due to a small negative charge

c) Electric field due to two small opposite charges

d) Electric field due to two small like charges

e) Uniform electric field between parallel charged plates

Figure 1.4 Electric field diagrams

In a uniform electric field, the lines of force are equally spaced and parallel to each other, showing that the force on a charged object is the same whatever its position in the field.

A battery has a positive terminal and a negative terminal. When a material is connected to these terminals there is an electric field, inside the material, between the positive and negative terminals. The electric field provides a force on the electrons, which could move from the negative terminal to the positive terminal. Charge could therefore be transferred through the material.

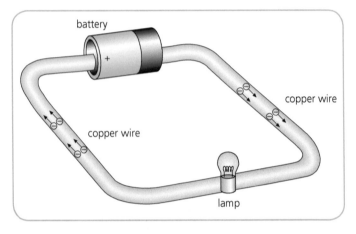

Figure 1.5 An electric current is the movement of electrons from the negative terminal to the positive terminal of a source of electrical energy, such as a battery

Physics beyond the classroom

Electrostatic painting uses charged paint droplets to efficiently paint curved metal objects. The paint spray is exposed to a high voltage as it leaves the spray gun. The tiny droplets of paint lose electrons and become positively charged. The object to be painted is given a negative charge. The positively charged paint particles are attracted to the negatively charged object. This method is more efficient than normal spray painting for two main reasons:

● The paint droplets spread out more when they leave the gun because they are positively charged and they repel each other; the paint covers a wider area.
● The paint droplets are attracted to the negatively charged object, so less paint is wasted on the floor and walls of the paint shop.

The lamp lights up. This is due to electrons from the negative terminal of the battery moving through the wires and lamp to the positive terminal of the battery. This movement of negative charges is called an **electric current** (or **current** for short). A current is a movement of electrons. Therefore, when there is a current, (negative) charge is transferred.

The amount of charge transferred is given by:

charge transferred = current × time of transfer
$$Q = It$$

where Q = charge transferred, measured in coulombs (C),
I = current, measured in amperes (A),
t = time of transfer, measured in seconds (s).

From the above:

$$I = \frac{Q}{t}$$

This means that 1 ampere = 1 coulomb per second ($1\,A = 1\,C\,s^{-1}$).

Charge, current and time

Consider a simple electrical circuit – a lamp connected to a battery – as shown in Figure 1.5.

Worked examples

Example 1

The current in a wire is 1.5 A. Calculate the charge transferred through the wire in 30 seconds.

Solution

$Q = It$

$Q = 1.5 \times 30$

$Q = 45 \, C$

Example 2

In a time of 5 minutes, 600 coulombs of charge pass through a lamp. Calculate the current in the lamp.

Solution

Note: t must be in seconds; 5 minutes = (5 × 60) s

$Q = It$

$600 = I \times (5 \times 60)$

$I = \dfrac{600}{300} = 2.0 \, A$

Example 3

When 1200 C of charge are transferred through a wire, the current in the wire is 0.5 A. Calculate the time taken for the charge to be transferred.

Solution

$Q = It$

$1200 = 0.5 \times t$

$t = \dfrac{1200}{0.5} = 2400 \, s$

Conductors and insulators

Electrons can only move from the negative terminal to the positive terminal of a battery if there is an electrical path between them. Materials that allow electrons to move through them easily, to form an electric current, are known as **conductors**. Conductors are mainly metals, such as copper, gold and silver. However, carbon, in the form of graphite, is also a good conductor.

Materials that do not allow electrons to move through them easily are called **insulators**. Glass, plastic, wood and air are examples of insulators.

Voltage or potential difference

A battery changes chemical energy into electrical energy. This electrical energy is carried by the electrons that move round the circuit. The electrical energy is converted into other forms of energy, for example heat and light, by components in the circuit such as a lamp. The amount of electrical energy the electrons have at any point in a circuit is called their 'potential'. As electrons move between two points in an electric circuit, they transfer electrical energy into other forms of energy. This means that the electrons have a different amount of electrical energy (or potential) at the two points. There is a **potential difference** (or **p.d.**) between the points. Potential difference (p.d.) is often referred to as **voltage**.

Physics beyond the classroom

Figure 1.6 This church spire has a lightning conductor

During certain weather conditions, clouds can store very large amounts of electrical charge. The charged clouds can discharge to the ground – a lightning strike – usually via tall buildings or trees. A lightning strike can cause considerable damage to a building. Because of this, tall buildings are fitted with lightning conductors. A lightning conductor consists of a thick copper strip connected to the ground. In the event of a lightning strike, the charge is conducted safely to the ground.

The voltage between the two terminals of the battery is a measure of the electrical energy given to the electrons by the battery. To be exact, the voltage of a battery is the electrical energy given to 1 coulomb of charge as it passes through the battery. For example, a 6.0 volt battery gives four times more electrical energy to each coulomb of charge passing through it compared with a 1.5 volt battery.

This means that 1 volt = 1 joule per coulomb ($1\,V = 1\,J\,C^{-1}$).

Circuit symbols

The circuit symbols for a cell, a battery, a resistor, a variable resistor, a switch, a lamp, an ammeter and a voltmeter are shown in Figure 1.7. Note that for the cell the 'taller' line represents the positive terminal and the 'shorter' line the negative terminal.

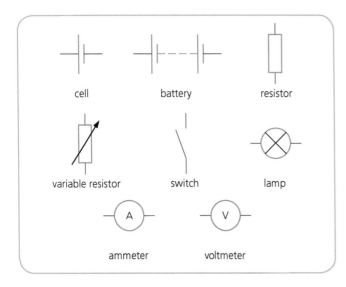

Figure 1.7 Common circuit symbols

Measuring current

An ammeter is used to measure electric current. Electric current is measured in amperes (A). Figure 1.8 shows how an ammeter is connected in an electrical circuit.

An ammeter measures the **current in a component** (it 'counts' the number of coulombs of charge passing a point each second).

Measuring voltage

A voltmeter is used to measure voltage or potential difference (p.d.). Potential difference is measured in volts (V). Figure 1.9 shows how a voltmeter is connected in an electrical circuit.

A voltmeter measures the **voltage or p.d. across a component** (the number of joules of energy transferred by each coulomb of charge).

Ammeters and voltmeters can be connected to the same circuit using the instructions given in Figures 1.8 and 1.9.

Resistance

All materials oppose current in them. This opposition to the current is called resistance. Resistance is measured in ohms (Ω).

For most materials resistance depends on:

● the type of material – the better the conductor the lower the resistance

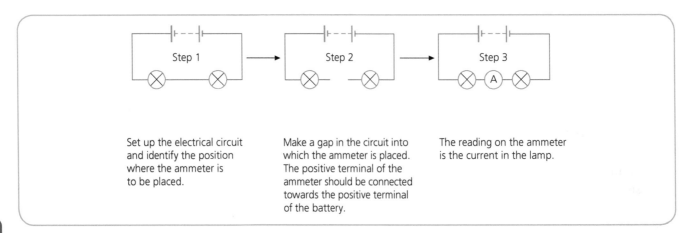

Figure 1.8 Connecting an ammeter into an electrical circuit

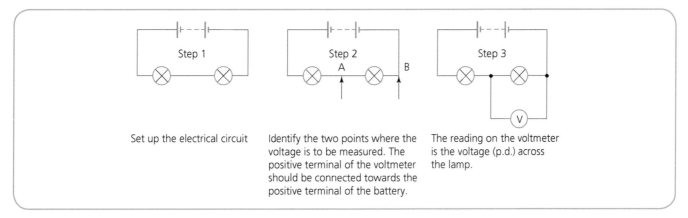

| Step 1 | Step 2 | Step 3 |

Set up the electrical circuit | Identify the two points where the voltage is to be measured. The positive terminal of the voltmeter should be connected towards the positive terminal of the battery. | The reading on the voltmeter is the voltage (p.d.) across the lamp.

Figure 1.9 Connecting a voltmeter into an electrical circuit

- the length of the material – the longer the material the higher the resistance
- the thickness of the material – the thinner the material the higher the resistance
- the temperature of the material – the higher the temperature the higher the resistance.

For a resistor the ratio:

$$\frac{\text{voltage (p.d.) across resistor}}{\text{current in resistor}}$$

remains approximately constant provided the temperature of the resistor remains constant. This constant is called the resistance of the resistor. Therefore, the resistance of a resistor can be calculated provided the voltage or potential difference across the resistor and the current in the resistor are known. Then:

$$\text{resistance} = \frac{\text{voltage (p.d.) across resistor}}{\text{current in resistor}}$$

i.e. $\qquad R = \dfrac{V}{I}$

This is Ohm's law. Ohm's law is usually written as:

$$\frac{\text{voltage (p.d.)}}{\text{across resistor}} = \frac{\text{current}}{\text{in resistor}} \times \frac{\text{resistance of}}{\text{resistor}}$$

$$V = IR$$

where V = voltage (or p.d.) across resistor, measured in volts (V),
I = current in resistor, measured in amperes (A),
R = resistance of resistor, measured in ohms (Ω).

The resistance of the resistor (in this case a lamp) shown in Figure 1.10 can be calculated using the voltmeter and ammeter readings.

$$\frac{\text{resistance}}{\text{of resistor}} = \frac{\text{voltage (p.d.) across resistor}}{\text{current in resistor}}$$

Figure 1.10 The resistance of a resistor (in this case a lamp) can be calculated using the voltmeter and ammeter readings

Worked examples

Example 1

Calculate the resistance of the lamp shown in Figure 1.10.

Solution

Reading on voltmeter = 1.44 V = voltage across the lamp
Reading on ammeter = 0.14 A = current in the lamp

$$V = IR$$
$$1.44 = 0.14 \times R$$
$$R = \frac{1.44}{0.14} = 10.3\,\Omega$$

Example 2

There is a current of 0.015 A in an LED (light emitting diode). The LED has a resistance of 120 Ω. Calculate the potential difference across the LED.

Solution

$$V = IR$$
$$V = 0.015 \times 120$$
$$V = 1.8\,V$$

Example 3

A kettle is connected to a 230 V supply and switched on. The element of the kettle has a resistance of 25 Ω. Calculate the current in the element.

Solution

$$V = IR$$
$$230 = I \times 25$$
$$I = \frac{230}{25} = 9.2\,A$$

Figure 1.11 shows how the current in a resistor, I, changes when the voltage across the resistor, V, is increased (the temperature of the resistor did not change). The graph of V against I a straight line through the origin, so V/I is constant, i.e. the resistance of the resistor is constant (see pages 172 and 173 for more information about graphs). (Use Ohm's law to calculate the resistance of the resistor for different voltages to check that the resistance of the resistor remains constant.)

Figure 1.12 shows how the current in a lamp, I, changes when the voltage across the lamp, V, is increased (the temperature of the lamp increases as the lamp gets brighter). In this case the graph of V against I is not a straight line through the origin and so the resistance of the lamp is not constant. (Use Ohm's law to calculate the resistance of the lamp at different voltages to check that the resistance of the lamp increases as the voltage increases.)

Note that the slope or gradient of a voltage–current graph gives the resistance.

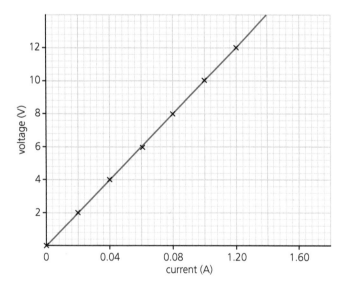

Figure 1.11 Graph of V against I for a resistor at constant temperature

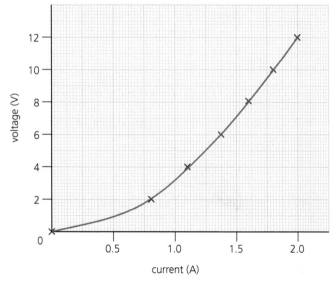

Figure 1.12 Graph of V against I for a filament lamp

A resistor whose resistance can be changed is known as a variable resistor. The resistance is changed by altering the length of the wire in the resistor; the longer the wire, the higher the resistance. The higher the resistance, the smaller the current. Variable resistors are often used as volume or brightness controls on televisions and as dimmers on lights.

Figure 1.13 Variable resistors

Physics beyond the classroom

At room temperature all materials have resistance. As most materials are cooled to lower temperatures, their resistance decreases. At very low temperatures of about −140 °C, the electrical resistance of some metals and alloys becomes negligible. These materials are known as superconductors.

Superconductors are ideal in applications where you do not want electrical energy changed into heat. This is put to use in strong electromagnets where an intense magnetic field is required.

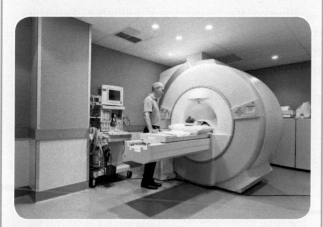

Figure 1.14 An MRI (magnetic resonance imaging) scanner uses a superconductor

Types of circuit

Electrical components, such as lamps and resistors, can be connected in **series**, in **parallel** or in a mixture of series and parallel. A series circuit has only one electrical path from the negative terminal of the battery to the positive terminal. A parallel circuit has more than one electrical path (called branches) from the negative terminal of the battery to the positive terminal. Figure 1.15 shows three lamps connected in series while Figure 1.16 shows three lamps connected in parallel. Figure 1.17 shows a mixed series and parallel circuit in which a lamp is connected in series with two resistors that are connected in parallel.

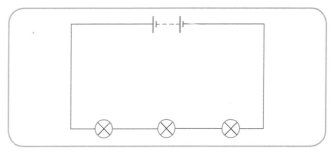

Figure 1.15 A series circuit

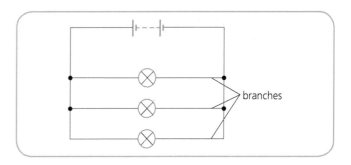

Figure 1.16 A parallel circuit

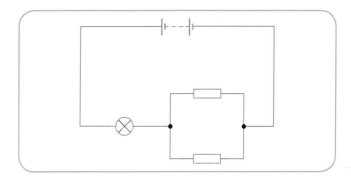

Figure 1.17 A mixed series and parallel circuit

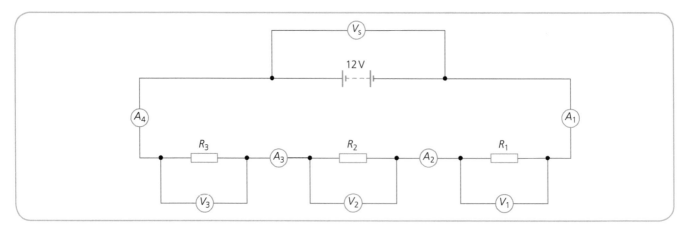

Figure 1.18 Measuring the current in, and the voltage (p.d.) across, resistors connected in series

A series circuit

A circuit is set up as shown in Figure 1.18. Ammeters are connected to measure the current at various positions. Voltmeters have also been connected to measure the voltage (p.d.) across each resistor and the battery. Table 1.1 shows the readings on the meters and the values of the resistances calculated using Ohm's law.

Current (A)	Voltage (V)	Resistance (Ω)
$A_1 = 0.15$	$V_1 = 1.5$	$R_1 = 10$
$A_2 = 0.15$	$V_2 = 6.0$	$R_2 = 40$
$A_3 = 0.15$	$V_3 = 4.5$	$R_3 = 30$
$A_4 = 0.15$	–	–
–	$V_{supply} = V_s = 12$	–

Table 1.1

From Table 1.1 we can conclude that:

- the current at different positions in a series circuit is the same
- the sum of the voltages (p.d.s) across the resistors is equal to the supply voltage.

Figure 1.19 shows a single resistor connected to the same supply voltage.

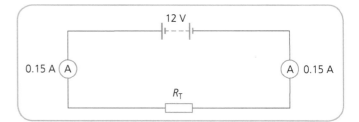

Figure 1.19 Equivalent circuit to that shown in Figure 1.18

This circuit must have the same combined or total resistance as that in Figure 1.18 since it has the same supply voltage and the same current. Using Ohm's law on this circuit gives a value for the single resistor of $80\,\Omega$, because:

$$R_T = \frac{V_{supply}}{I} = \frac{12}{0.15} = 80\,\Omega$$

Comparing Figures 1.18 and 1.19 we see that the combined or total resistance of a number of resistors connected in series is equal to the sum of the individual resistances. Adding resistors in series increases the total resistance of the circuit.

For the series circuit shown in Figure 1.20:

- the current is the same at all positions – current does not split up
- the supply voltage is equal to the sum of the voltages (potential differences) across the components, i.e. $V_{supply} = V_1 + V_2 + V_3$
- the total resistance (R_T) of the circuit is found using $R_T = R_1 + R_2 + R_3$.

Note: for a series circuit the total resistance is greater than the value of the largest resistance connected in series.

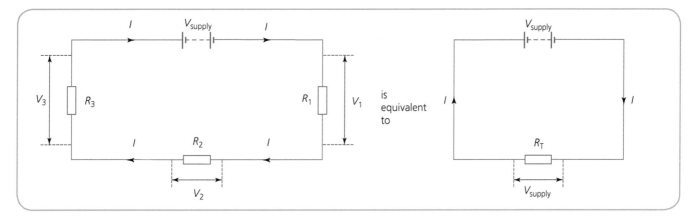

Figure 1.20 Equivalent circuits provided $R_T = R_1 + R_2 + R_3$

Worked examples

Example 1

Three resistors of value $100\,\Omega$, $47\,\Omega$, and $33\,\Omega$ are connected in series. What is the total resistance of the resistors?

Solution

$R_T = R_1 + R_2 + R_3$
$R_T = 100 + 47 + 33$
$R_T = 180\,\Omega$

Example 2

A circuit is set up as shown in Figure 1.21. Calculate the ammeter readings A_1 and A_2 and the potential difference across each of the resistors.

Solution

total resistance of circuit, $R_T = R_1 + R_2 + R_3 + R_4$
$R_T = 4 + 8 + 10 + 2$
$R_T = 24\,\Omega$

From Ohm's law:

circuit current $= I = \dfrac{V_{supply}}{R_T} = \dfrac{12}{24} = 0.5\,A$

$A_1 = A_2 = 0.5\,A$ (since current in a series circuit is the same at all points)

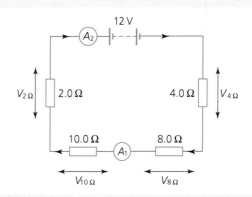

Figure 1.21

The potential differences across each resistor are calculated using Ohm's law:

$$\text{voltage (p.d.) across resistor} = \text{current in resistor} \times \text{resistance of resistor}$$

voltage across $4\,\Omega$ resistor $= V_{4\Omega} = (IR)_{4\Omega} = 0.5 \times 4 = 2.0\,V$
voltage across $8\,\Omega$ resistor $= V_{8\Omega} = (IR)_{8\Omega} = 0.5 \times 8 = 4.0\,V$
voltage across $10\,\Omega$ resistor $= V_{10\Omega} = (IR)_{10\Omega} = 0.5 \times 10 = 5.0\,V$
voltage across $2\,\Omega$ resistor $= V_{2\Omega} = (IR)_{2\Omega} = 0.5 \times 2 = 1.0\,V$

A parallel circuit

A circuit is set up as shown in Figure 1.22. Ammeters are connected to measure the current at various positions. Voltmeters have also been connected to measure the voltage (p.d.) across each resistor and the battery. Table 1.2 shows the readings on the meters and the values of the resistances calculated using Ohm's law.

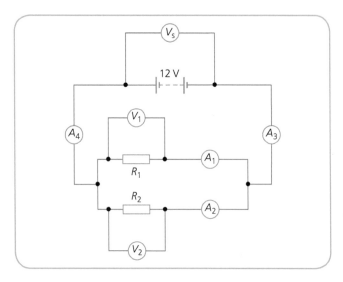

Figure 1.22 Measuring the current in, and the voltage (p.d.) across, resistors connected in parallel

Current (A)	Voltage (V)	Resistance (Ω)
$A_1 = 0.2$	$V_1 = 12$	$R_1 = 60$
$A_2 = 0.1$	$V_2 = 12$	$R_2 = 120$
$A_3 = 0.3$	–	–
$A_4 = 0.3$	–	–
–	$V_{supply} = V_s = 12$	–

Table 1.2

From Table 1.2 we can conclude that:

- the current from the supply is equal to the sum of the currents in the branches
- the voltages (p.d.s) across the resistors connected in parallel are the same.

Figure 1.23 shows a single resistor connected to the same supply voltage. This circuit must have the same combined or total resistance as that in Figure 1.22 since it has the same supply voltage and the same current from the supply. Using Ohm's law on this circuit gives a value for the single resistor of $40\,\Omega$, because:

$$R_T = \frac{V_{supply}}{I} = \frac{12}{0.3} = 40\,\Omega$$

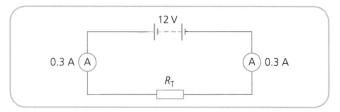

Figure 1.23 Equivalent circuit to that shown in Figure 1.22

Comparing Figures 1.22 and 1.23 we see that the combined resistance of a number of resistors connected in parallel is smaller than any of the individual resistances.

The combined resistance of $40\,\Omega$ is obtained for these resistors as follows:

$$\frac{1}{R_T} = \frac{1}{R_1} + \frac{1}{R_2} = \frac{1}{60} + \frac{1}{120} = 0.017 + 0.008 = 0.025$$

$$\frac{1}{R_T} = 0.025$$

$$R_T = \frac{1}{0.025} = 40\,\Omega$$

Adding resistors in parallel reduces the resistance of that part of the circuit.

For the parallel circuit shown in Figure 1.24 we have:

- Current from supply = sum of currents in the branches, i.e. $I = I_1 + I_2$
- The voltage (p.d.) across resistors is the same, i.e. $V_1 = V_2$ and in this case $V_{supply} = V_1 = V_2$
- The total resistance of the circuit is found using

$$\frac{1}{R_T} = \frac{1}{R_1} + \frac{1}{R_2}$$

Note: for a parallel circuit, the total resistance is less than the value of the smallest resistance connected in parallel.

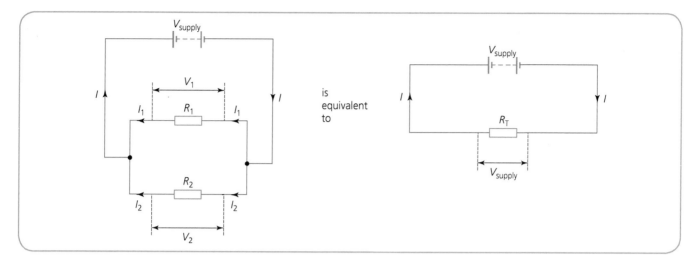

Figure 1.24 Equivalent circuits provided $\dfrac{1}{R_T} = \dfrac{1}{R_1} + \dfrac{1}{R_2}$

Worked examples

Example 1

Three resistors of resistance $20\,\Omega$, $60\,\Omega$ and $30\,\Omega$ are connected in parallel as shown in Figure 1.25. Calculate the resistance between X and Y.

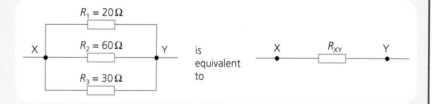

Figure 1.25 A parallel circuit

Solution

$$\frac{1}{R_{XY}} = \frac{1}{R_1} + \frac{1}{R_2} + \frac{1}{R_3} = \frac{1}{20} + \frac{1}{60} + \frac{1}{30}$$

$$\frac{1}{R_{XY}} = 0.050 + 0.017 + 0.033 = 0.100$$

$$R_{XY} = \frac{1}{0.100} = 10\,\Omega$$

As a check on the answer we would expect R_{XY} to be smaller than $20\,\Omega$.

Example 2

Two resistors each of resistance $1.0\,\text{k}\Omega$ are connected in parallel. Calculate their total resistance.

Solution

Note: $1\,\text{k}\Omega = 1 \times 10^3\,\Omega = 1000\,\Omega$

$$\frac{1}{R_T} = \frac{1}{R_1} + \frac{1}{R_2} = \frac{1}{1000} + \frac{1}{1000} = 0.001 + 0.001$$

$$\frac{1}{R_T} = 0.002$$

$$R_T = \frac{1}{0.002} = 500\,\Omega$$

Note that the total resistance of two identical resistors connected in parallel is half that of one of the resistors.

Example 3

A circuit is set up as shown in Figure 1.26. Calculate
a) the total resistance of the circuit,
b) the current drawn from the battery,
c) the p.d. across the $10\,\Omega$ resistor.

Solution

a) $$\frac{1}{R_{XY}} = \frac{1}{R_1} + \frac{1}{R_2} = \frac{1}{100} + \frac{1}{25} = 0.010 + 0.040 = 0.050$$

$$R_{XY} = \frac{1}{0.050} = 20\,\Omega$$

total resistance $R_T = 10 + R_{XY} = 10 + 20 = 30\,\Omega$

b) $V_{supply} = IR_T$

$12 = I \times 30$

$I = \dfrac{12}{30} = 0.4\,A$

c) $\begin{array}{ccccc} \text{voltage across} & = & \text{current in} & \times & \text{resistance of} \\ 10\,\Omega \text{ resistor} & & 10\,\Omega \text{ resistor} & & 10\,\Omega \text{ resistor} \end{array}$

$V = IR$

$V = 0.4 \times 10$

$V = 4.0\,V$

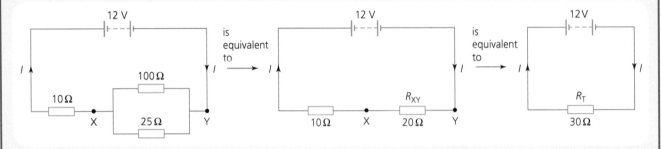

Figure 1.26 Example of a mixed series and parallel circuit

Key facts and physics equations: electrical circuits

- There are two types of charge: positive charge and negative charge.
- Like charges repel, unlike charges attract.
- Charge transferred = current × time, i.e. $Q = It$.
- Charge is measured in coulombs (C), current in amperes (A) and time in seconds (s).
- The voltage of a supply is the energy given to 1 coulomb of charge as it passes through the supply.
- An ammeter is connected in series with the component.
- A voltmeter is connected in parallel across the component.
- Voltage (p.d.) across a resistor = current in resistor × resistance of resistor, i.e. $V = IR$. This is Ohm's law.
- Voltage (or p.d.) is measured in volts (V), current in amperes (A) and resistance in ohms (Ω).
- The resistance of a resistor remains constant for different currents provided the temperature of the resistor does not change.

- In a series circuit (see Figure 1.27):
  - the current is the same at all points, i.e. $I_1 = I_2 = I_3$
  - the supply voltage is equal to the sum of the voltages (p.d.s) across components, i.e. $V_{supply} = V_1 + V_2 + V_3$

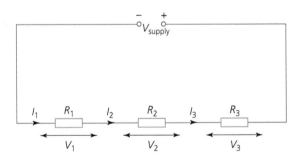

Figure 1.27 A series circuit

  - the total (or combined) resistance (R_T) is found using $R_T = R_1 + R_2 + R_3$
  - the total resistance is greater than the value of the largest resistance connected in series.

- In a parallel circuit (see Figure 1.28):
  - the circuit current from the supply is equal to the sum of the currents in the branches, i.e. $I_{circuit} = I_1 + I_2 + I_3$
  - the voltage (p.d.) across components is the same, i.e. $V_{supply} = V_1 = V_2 = V_3$
  - the total (or combined) resistance (R_T) is found using $\dfrac{1}{R_T} = \dfrac{1}{R_1} + \dfrac{1}{R_2} + \dfrac{1}{R_3}$
  - the total resistance is less than the value of the smallest resistance connected in parallel.

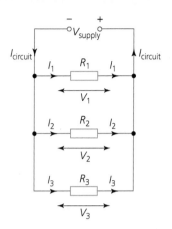

Figure 1.28 A parallel circuit

End-of-chapter questions

1 A hairdryer is switched on for 3 minutes. The current in the element of the hairdryer is 5.0 amperes. Calculate the charge transferred by the hairdryer element in this time.

2 The current in a lamp is 2.0 A.
a) State what is meant by the term *current*.
b) The lamp transfers 800 C of charge. Calculate the time taken to transfer this charge.

3 A charge of 750 C passes through a resistor in 5 minutes. Calculate the current in the resistor.

4 An electronic game works from a 9.0 V supply. What does a supply voltage of 9.0 V mean?

5 Draw the circuit symbol for:
a) a battery
b) a lamp
c) a switch
d) a resistor
e) a variable resistor
f) an ammeter
g) a voltmeter.

6 Circuits are set up as shown in Figure 1.29. What are the readings on a) ammeters A_1, A_2 and A_3, b) voltmeters V_1, V_2 and V_3?

7 Three resistors of value 47 Ω, 100 Ω and 150 Ω are connected in series. Calculate the total resistance of these three resistors.

8 Three resistors of value 20 Ω, 20 Ω and 10 Ω are connected in parallel. Calculate the total resistance of these three resistors.

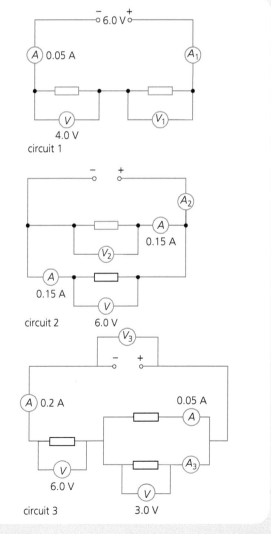

Figure 1.29

9 Four resistors are arranged as shown in Figure 1.30. Calculate the resistance between X and Y.

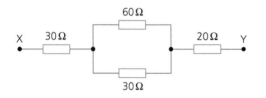

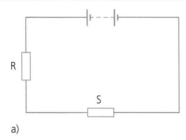

Figure 1.30

10 Circuits are set up as shown in Figure 1.31. Redraw each of the circuit diagrams to show how **both** a voltmeter is connected to measure the voltage across component R **and** an ammeter is connected to measure the current in component S.

11 The current in a 1.0 kΩ resistor is 0.012 A. Calculate the potential difference across the resistor.

12 An electric toaster is connected to the 230 V mains and switched on. There is a current of 4.5 A in the toaster element. Calculate the resistance of the element.

13 The motor of a toy car is connected to a 4.5 V battery. The motor of the car has a resistance of 18 Ω. Calculate the current in the motor.

14 Two resistors, of value 1.0 kΩ and 3.0 kΩ, are connected in series with a 9.0 V battery. What is the potential difference across the 3.0 kΩ resistor?

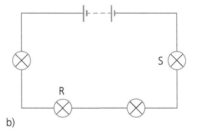

a)   b)

Figure 1.31

2 Electrical energy and power

Learning outcomes

At the end of this chapter you should be able to:

1 State that when there is an electrical current in a component, there is an energy transformation.
2 State that in a lamp electrical energy is transformed into heat and light.
3 State that the energy transformation in an electrical heater occurs in the resistance wire.
4 State that power is the electrical energy transferred each second.
5 State that the electrical energy transformed each second = VI.

6 Carry out calculations involving the relationships between power, energy, time, current and potential difference.
7 Explain the equivalence between IV, I^2R and V^2/R.
8 Carry out calculations involving the relationships between power, current, voltage and resistance.
9 Draw and identify the circuit symbol for a fuse.
10 Select an appropriate fuse given the power rating of an appliance.
11 State that a battery is d.c. and the mains supply is a.c.
12 Explain both d.c. and a.c. in terms of current.

Power

When there is an electric current in a wire, the electrons making up the current collide with the atoms of the wire. These collisions make the atoms vibrate more and this results in the wire becoming hotter, i.e. electrical energy has been changed into heat in the wire. The amount of heat produced depends on the value of the current and the value of the resistance of the wire.

Heating elements for electric kettles and hairdryers change electrical energy into heat in the wire making up the element.

A lamp transfers electrical energy into heat and light in a wire called the filament.

Power is the energy transferred in 1 second.

$$\text{power} = \frac{\text{energy transferred}}{\text{time taken}}$$
$$P = \frac{E}{t}$$

where P = power, measured in watts (W),
E = energy transferred, measured in joules (J),
t = time taken, measured in seconds (s).

Therefore 1 watt = 1 joule per second (1 W = 1 J s^{-1}).

Worked examples

Example 1

A hairdryer uses 86.4 kJ of energy in a time of 1.5 minutes. Calculate the power rating of the hairdryer.

Solution

Note: 86.4 kJ = 86.4 × 10^3 J = 86 400 J
and 1.5 minutes = (1.5 × 60) s

$$P = \frac{E}{t} = \frac{86.4 \times 10^3}{1.5 \times 60} = 960 \text{ W}$$

Example 2

The power rating of the element of an electric kettle is 2.2 kW. The element uses 176 kJ of energy to heat some water. Calculate the time the element is switched on.

Solution

Note: 2.2 kW = 2.2 × 10^3 W = 2200 W
and 176 kJ = 176 × 10^3 J = 176 000 J

$$P = \frac{E}{t}$$

$$2.2 \times 10^3 = \frac{176 \times 10^3}{t}$$

$$t = \frac{176 \times 10^3}{2.2 \times 10^3} = 80\,s$$

Example 3

A colour television set is rated at 280 W. The television is switched on for 5 hours. Calculate the energy the television will use in this time.

Solution

Note: 5 hours = (5 × 60) minutes = (5 × 60 × 60) s

$$P = \frac{E}{t}$$

$$280 = \frac{E}{(5 \times 60 \times 60)}$$

$$E = 280 \times 18\,000 = 5.04 \times 10^6\,J$$

Power, current and voltage

Three different lamps of known power ratings are connected to an electrical supply. The current in and the voltage (p.d.) across each lamp is recorded. The readings are shown in Table 2.1.

Lamp	Power rating of lamp (W)	Current in lamp (A)	Voltage across lamp (V)
X	24	2	12
Y	36	3	12
Z	48	4	12

Table 2.1

From Table 2.1 we can conclude:

power = current × voltage

$P = IV$

But from Ohm's law: $V = IR$

$P = IV = I \times (IR)$

$P = I^2R$

Alternatively:

$P = IV$

But from Ohm's law: $I = \dfrac{V}{R}$

$$P = IV = \frac{V}{R} \times V$$

$$P = \frac{V^2}{R}$$

The equations $P = IV$, $P = I^2R$ and $P = \dfrac{V^2}{R}$ can be used to find the power rating of appliances.

Use Ohm's law to calculate the resistance of each of the lamps X, Y and Z in the table and then check that the above equations can be used to calculate the power rating of the lamps.

Worked examples

Example 1

The element of an electric heater is plugged into the 230 V mains supply. The heater is switched on and there is a current of 4.5 A in the element. Calculate the power rating of the element.

Solution

$P = IV = 4.5 \times 230 = 1035\,W$

Example 2

When operating, the element of an electric kettle has a resistance of 23 Ω. The current in the element is 10 A. Calculate the power rating of the element.

Solution

$P = I^2R = 10^2 \times 23 = 2300\,W$

Example 3

The element of an electric toaster has a power rating of 1050 W. The toaster is connected to a 230 V mains supply and switched on. Calculate the resistance of the element.

Solution

$$P = \frac{V^2}{R}$$

$$1050 = \frac{230^2}{R}$$

$$R = \frac{230^2}{1050} = 50.4\,\Omega$$

Power and plug fuses

When an electric current passes through a wire, some of the electrical energy is changed into heat. If too large a current passes through the wire then the wire could overheat and catch fire. To prevent this, (three-pin) plugs are fitted with a safety device called a **fuse**. When the value of the current is greater than the fuse value, the fuse wire becomes so hot that it melts or 'blows', breaking the electrical circuit it is part of.

Figure 2.1 shows the circuit symbol for a fuse. Fuses are available in a number of values but the most common are 3 A and 13 A. Figure 2.2 shows a three-pin plug fitted with 3 A fuse.

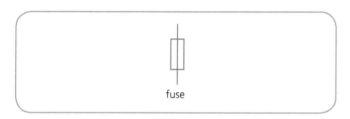

Figure 2.1 Circuit symbol for a fuse

Figure 2.2 A three-pin plug fitted with a 3 A fuse

The value of fuse that is fitted is mainly dependent on the power rating of the appliance. In general, if the power rating of an appliance is:

- less than 700 W then a 3 A fuse should be fitted
- greater than or equal to 700 W then a 13 A fuse should be fitted to the plug.

If a more detailed value is required then the current in a working appliance can be calculated using the equation $I = \dfrac{P}{V}$. The next highest fuse value would then be chosen for this current.

Example

A microwave oven is to operate from the 230 V mains supply. The power rating of the microwave is 800 W. The plug of the microwave is to be fitted with a fuse rated at either 3 A or 13 A. Determine, by calculation, the rating of the fuse fitted in the plug of the microwave.

Solution

$P = IV$

$800 = I \times 230$

$I = \dfrac{800}{230} = 3.48\,\text{A}$

The circuit current is greater than 3 A hence the (next highest) fuse is 13 A.

Direct and alternating current

Figure 2.3 shows a battery connected to a lamp. Figure 2.4 shows a mains-operated, low-voltage power supply connected to an identical lamp. The lamps are equally bright.

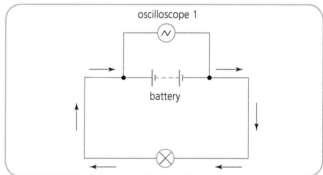

Figure 2.3 Electrons only move in one direction. This is known as direct current (d.c.)

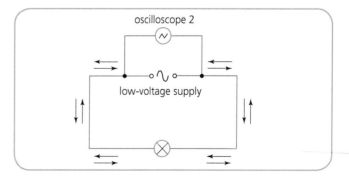

Figure 2.4 Electron movement is to and fro. This is known as alternating current (a.c.)

In Figure 2.3 electrons (negative charges) move from the negative terminal through the lamp and wires to the positive terminal of the battery. This means that the electrons move in only one direction – this is known as direct current or d.c.

In Figure 2.4 electrons move in one direction, then in the other direction and back again, i.e. the electrons move to and fro. This alternating movement of the electrons is known as alternating current or a.c. The to and fro movement of the electrons is frequent – it occurs 50 times every second. This is called the frequency of the mains and is stated as 50 hertz.

Figure 2.5 shows the trace observed on the screen of oscilloscope 1. The trace has a constant value of 1.5 V. This is called d.c.

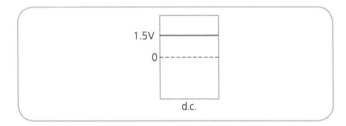

Figure 2.5 Oscilloscope trace from Figure 2.3

Figure 2.6 shows the trace observed on the screen of oscilloscope 2. The trace alternates from a maximum or peak value of +2.1 V to −2.1 V. This is called a.c.

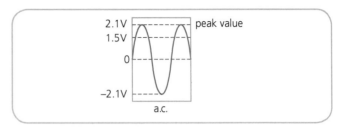

Figure 2.6 Oscilloscope trace from Figure 2.4

Key facts and physics equations: electrical energy and power

- Power = $\dfrac{\text{energy}}{\text{time}}$, i.e. $P = \dfrac{E}{t}$.
- Power = current × voltage, i.e. $P = IV$.
- Power = current^2 × resistance, i.e. $P = I^2R$.
- Power = $\dfrac{\text{voltage}^2}{\text{resistance}}$, i.e. $P = \dfrac{V^2}{R}$.
- Power is measured in watts (W), energy in joules (J), time in seconds (s), current in amperes (A), voltage in volts (V) and resistance in ohms (Ω).
- Appliances with a power rating less than 700 W are normally fitted with a 3 A fuse.
- Appliances with a power rating greater than or equal to 700 W are fitted with a 13 A fuse.
- Direct current (d.c.) – electrons move in only one direction round a circuit.
- Alternating current (a.c.) – electrons move to and fro in a circuit.

End-of-chapter questions

1. The element of an electric iron is connected to the 230 V mains supply and switched on. The current in the element is 5.5 A.
   a) Calculate the power rating of the element.
   b) How much electrical energy does the element use in 1 second?

2. A filament lamp is plugged into a 12 V supply and switched on. When lit, the resistance of the lamp is 3.0 Ω.
   a) Write down the energy change which takes place in the filament of the lamp.
   b) Calculate the current in the lamp.
   c) Calculate the power rating of the lamp.

3. An electric oven is operating from the 230 V mains supply. The resistance of the element of the oven is 17.6 Ω. Calculate the power rating of the element.

4. A spotlight is rated at 12 V, 50 W. The spotlight is switched on and is operating at its rated values. Calculate the resistance of the spotlight.

5. The power rating of a motor is 138 W.
   a) What is meant by *power rating of a motor is 138 W*?
   b) The current in the electric motor is 0.6 A. Calculate the resistance of the motor when it is operating.

6. a) Draw the circuit symbol for a fuse.
   b) What is the purpose of a fuse?

7. What value of fuse, 3 A or 13 A, should be fitted to the three-pin plug of a:

   a) 60 W table lamp
   b) 1400 W vacuum cleaner
   c) 1100 W toaster?

8. A circuit is set up as shown below.

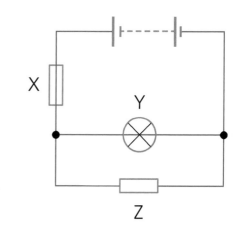

Figure 2.7

Name components X, Y and Z.

9. Explain in terms of electron flow in a circuit what is meant by:
   a) direct current (d.c.)
   b) alternating current (a.c.).

3 Electrical components and electronic circuits

Input devices

Microphone

A microphone is connected to an oscilloscope as shown in Figure 3.1. The amplitude of the trace displayed on the oscilloscope screen increases as louder notes are played into the microphone. A microphone changes sound into electrical energy. The louder the sound, the greater the electrical energy produced.

Figure 3.1 A microphone connected to an oscilloscope

Thermocouple

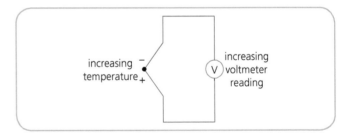

Figure 3.2 A thermocouple connected to a voltmeter

A thermocouple is composed of two different types of wire joined together. A thermocouple is connected to a voltmeter as shown in Figure 3.2. When the junction of the thermocouple is placed in a Bunsen flame, the voltmeter reading increases. A thermocouple changes heat into electrical energy. The higher the temperature of the junction, the greater the electrical energy produced.

Solar cell

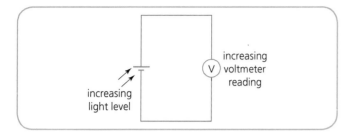

Figure 3.3 A solar cell connected to a voltmeter

A solar cell (or photovoltaic cell) is connected to a voltmeter as shown in Figure 3.3. When the solar cell is exposed to more light, the voltmeter reading increases. A solar cell changes light into electrical energy. The brighter the light shining on the solar cell, the greater the electrical energy produced.

Thermistor

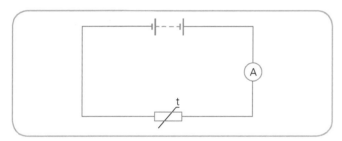

Figure 3.4 A thermistor circuit

When the thermistor, shown in Figure 3.4, is heated, the ammeter reading increases – therefore the resistance of the thermistor must be decreasing.

The resistance of most thermistors usually decreases with increasing temperature.

temperature ↑, resistance of thermistor ↓

Example

A circuit is set up as shown in Figure 3.5.

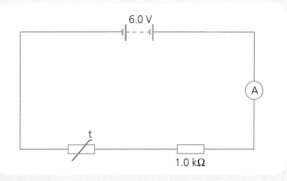

Figure 3.5

The thermistor is at a temperature of 20 °C. The reading on the ammeter is 2.0 mA.

a) Calculate the voltage across the 1.0 kΩ resistor at 20 °C.

b) Find the resistance of the thermistor at 20 °C.

Solution

a) Note: $2.0 \text{ mA} = 2 \times 10^{-3} \text{ A} = 0.002 \text{ A}$
   and $1.0 \text{ k}\Omega = 1 \times 10^3 \Omega = 1000 \Omega$

$$V_{resistor} = (IR)_{resistor} = 2 \times 10^{-3} \times 1 \times 10^3 = 2.0 \text{ V}$$

b) $V_{supply} = V_{resistor} + V_{thermistor}$

$6 = 2 + V_{thermistor}$

$V_{thermistor} = 4.0 \text{ V}$

But $V_{thermistor} = (IR)_{thermistor}$

$4 = 2 \times 10^{-3} \times R_{thermistor}$

$$R_{thermistor} = \frac{4}{(2 \times 10^{-3})} = 2000 \Omega$$

Light-dependent resistor (LDR)

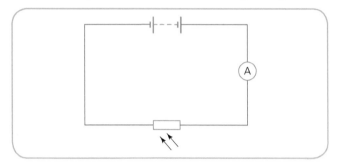

Figure 3.6 An LDR circuit

When the light-dependent resistor, shown in Figure 3.6, is exposed to brighter light, the ammeter reading increases – therefore the resistance of the LDR must be decreasing.

As the light gets brighter (light intensity increases), the resistance of the LDR decreases.

light intensity ↑, resistance of LDR ↓

Worked example

Example

A circuit is set up as shown in Figure 3.7.

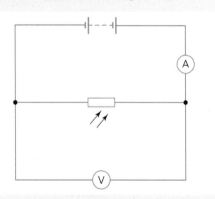

Figure 3.7

Light is shone on the LDR. The reading on the voltmeter is 2.0 V. The reading on the ammeter is 8.0 mA. Calculate the resistance of the LDR at this light level.

Solution

Note: $8.0 \text{ mA} = 8 \times 10^{-3} \text{ A} = 0.008 \text{ A}$

$$V_{LDR} = (IR)_{LDR}$$

$$2 = 8 \times 10^{-3} \times R_{LDR}$$

$$R_{LDR} = \frac{2}{(8 \times 10^{-3})} = 250\,\Omega$$

Switch

A d.c. supply is connected to a lamp and a switch as shown in Figure 3.8.

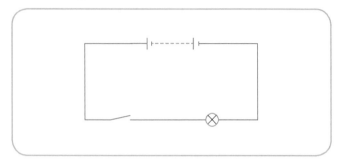

Figure 3.8 A voltage supply connected to a switch and a lamp

When the switch is open the lamp does not light. This is due to there being an air gap between the metal contacts of the switch (resistance is very, very high) – the resistance of an open switch is infinite.

When the switch is closed the lamp lights. This is due to the metal contacts of the switch touching (resistance is very, very low) – the resistance of a closed switch is zero.

Capacitor

A capacitor consists of two metal plates separated by an insulator. The construction of a capacitor is shown in Figure 3.9. It is a device that can store electric charge. The units of capacitance are farads (F). Most capacitors have very small values and so are measured in microfarads (μF), i.e. millionths of a farad.

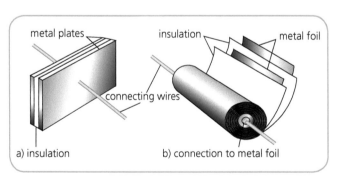

a) insulation   b) connection to metal foil

Figure 3.9 a) A capacitor consists of two metal plates separated by an insulator. **b)** Practical capacitors are constructed like a 'Swiss roll'

Figure 3.10a shows a capacitor with no charge on its plates. Figure 3.10b shows the capacitor when it is fully charged. This occurs when the voltage across the plates of the capacitor is equal to the voltage available for the capacitor – in this case the supply voltage. Figure 3.10c shows how a charged capacitor can be discharged by connecting the two plates together using a switch. When the switch is closed, the capacitor is discharged.

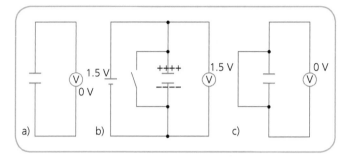

Figure 3.10 a) Uncharged capacitor, **b)** fully charged capacitor, **c)** discharged capacitor (when switch is closed)

A discharged capacitor has no voltage across its plates. The voltage across a fully charged capacitor is equal to the voltage available for the capacitor.

Output devices
Loudspeaker

Figure 3.11 A signal generator connected to a loudspeaker

A loudspeaker is connected to a signal generator as shown in Figure 3.11. As the amplitude (energy of the electrical signal) from the signal generator is increased, the sound from the loudspeaker gets louder. A loudspeaker changes electrical energy into sound.

Electric motor

Figure 3.12 A variable voltage supply connected to an electric motor

An electric motor is connected to a variable d.c. supply voltage as shown in Figure 3.12. The speed of the motor increases as the voltage (p.d.) is increased. An electric motor changes electrical energy into kinetic energy. Reversing the connections to the supply reverses the direction of rotation of the motor.

Relay

A relay is a switch operated by an electromagnet. A coil of wire when carrying an electric current provides the magnetic field required to close the relay switch shown in Figure 3.13.

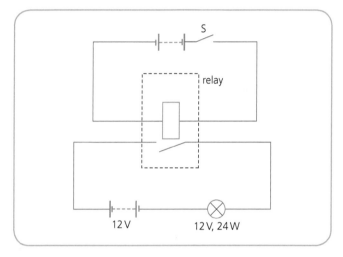

Figure 3.13 A relay circuit – closing switch S allows the lamp to light

When switch S is closed there is a current in the relay coil surrounding the relay switch. The magnetic field produced pulls the relay switch closed. This completes the lower electrical circuit and the lamp lights. When switch S is opened there is no current in the relay coil (and no magnetic field to pull the relay switch closed), the relay switch opens and the lamp goes out. The relay changes electrical energy into the opening or closing of a switch.

Filament lamp

Figure 3.14 A variable voltage supply connected to a lamp

A filament lamp consists of a thin tungsten wire (filament) in a glass container. When there is an electric current in the wire, electrical energy is changed into heat and light in the filament.

A lamp is connected to a variable d.c. voltage supply as shown in Figure 3.14. Increasing the voltage (p.d.) across the lamp increases the current in the filament and so the lamp gets brighter. No difference is observed when the connections from the supply to the lamp are reversed. The filament in the lamp requires a relatively large current to light properly and gets very hot in operation.

Light-emitting diode (LED)

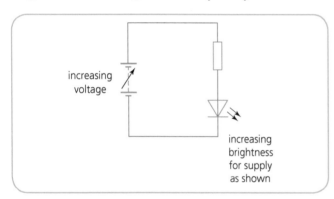

Figure 3.15 A variable voltage supply connected to a resistor and an LED

Light-emitting diodes are made by joining two special materials (called **semiconductors**) together to produce a junction. When there is an electric current in the junction, electrical energy is changed into light. However, too large a current – or indeed too high a p.d. – will destroy the junction, To prevent this, a resistor must be connected in series with the LED. Figure 3.15 shows an LED and a resistor connected to a variable d.c. voltage supply. Increasing the voltage (p.d.) across the LED increases its brightness. The LED does not light when the connections from the supply are reversed. The LED only requires a small current to light and does not get hot in operation. LEDs are available that emit red, green, yellow, blue and white light.

(Note that electrons can only move in the opposite direction from the arrow shown on the LED symbol.)

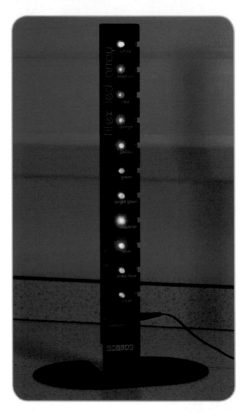

Figure 3.16 LEDs emit a number of different colours

Figure 3.17 Various filament lamps

Worked example

Example

A circuit is set up as shown in Figure 3.18.

The maximum voltage allowed across the LED is 1.8 V. The current in the LED must not exceed 10 mA. Calculate the resistance of the resistor, *R*.

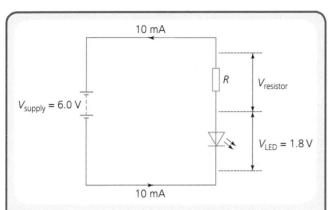

Figure 3.18

Solution

Note: $10\,\text{mA} = 10 \times 10^{-3}\,\text{A} = 0.010\,\text{A}$

Since the LED and resistor are connected in series then $V_{\text{supply}} = V_{\text{LED}} + V_{\text{resistor}}$ and the current in both components is the same. Therefore:

$$V_{\text{resistor}} = V_{\text{supply}} - V_{\text{LED}} = 6.0 - 1.8 = 4.2\,\text{V}$$

$$V_{\text{resistor}} = (IR)_{\text{resistor}}$$

$$4.2 = 10 \times 10^{-3} \times R$$

$$R = \frac{4.2}{(10 \times 10^{-3})} = 420\,\Omega$$

Physics beyond the classroom

Light-emitting diode (LED) lamps are replacing filament lamps in traffic lights and side and tail lamps for some buses and cars. LED lamps consist of a number of ultra-bright LEDs. LED lamps have much lower power consumption, longer life and are more reliable and brighter than filament lamps.

Figure 3.19 LED lamps are used to light this set of traffic lights

Another device
Diode

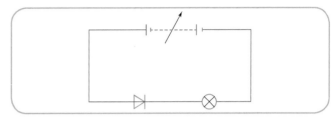

Figure 3.20 A variable voltage supply connected to a diode and a lamp

A diode is similar in construction to an LED, as it is made by joining two semiconductor materials together to produce a junction. Figure 3.20 shows a diode and a lamp connected to a variable d.c. voltage supply. As the supply voltage (p.d.) is increased, the lamp brightness increases. The lamp does not light when the connections to the supply are reversed.

A diode only conducts when connected to a d.c. supply the correct way round.

Diodes and LEDs are similar devices except that an LED gives out light. A diode can be used to prevent damage to some types of capacitor by incorrect connection of a d.c. supply. These devices are called **polarity sensitive devices**.

(Note: electrons can only move in the opposite direction from the arrow shown on the diode symbol.)

Resistors in series – a voltage divider

For a series circuit, the supply voltage is equal to the sum of the potential differences (voltages) across the individual resistors, i.e. $V_{\text{supply}} = V_1 + V_2$. This means that the supply voltage is split up, in this case, into two smaller bits and so the supply voltage is divided between the two resistors. Consider the circuit set up in Figure 3.21.

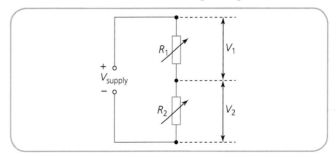

Figure 3.21 A voltage divider circuit

Let $R_1 = 1.0\,k\Omega$, $R_2 = 1.0\,k\Omega$ and $V_{supply} = 6.0\,V$

$R_T = R_1 + R_2 = 1000 + 1000 = 2000\,\Omega$

Note: $1.0\,k\Omega = 1 \times 10^3\,\Omega = 1000\,\Omega$

$V_{supply} = I_{circuit} \times R_T$

$6 = I_{circuit} \times 2000$

$I_{circuit} = \dfrac{6}{2000} = 0.003\,A$

$V_1 = I_{circuit} \times R_1 = 0.003 \times 1000 = 3.0\,V$

$V_2 = I_{circuit} \times R_2 = 0.003 \times 1000 = 3.0\,V$

$V_{supply} = V_1 + V_2 = 6\,V$

In Figure 3.21, let $R_1 = 2.0\,k\Omega$ (resistance of R_1 is increased to $2.0\,k\Omega$), $R_2 = 1.0\,k\Omega$ and $V_{supply} = 6.0\,V$.

$R_T = R_1 + R_2 = 2000 + 1000 = 3000\,\Omega$

$V_{supply} = I_{circuit} \times R_T$

$6 = I_{circuit} \times 3000$

$I_{circuit} = \dfrac{6}{3000} = 0.002\,A$

$V_1 = I_{circuit} \times R_1 = 0.002 \times 2000 = 4.0\,V$

$V_2 = I_{circuit} \times R_2 = 0.002 \times 1000 = 2.0\,V$

$V_{supply} = V_1 + V_2 = 6\,V$

As the resistance of R_1 increases, the voltage across R_1 (V_1) increases. This means that the voltage across R_2 (V_2) decreases, although the resistance of R_2 did not change ($V_{supply} = V_1 + V_2 = 6.0\,V$).

In Figure 3.21, let $R_1 = 1.0\,k\Omega$, $R_2 = 5.0\,k\Omega$ (resistance of R_2 is increased to $5.0\,k\Omega$) and $V_{supply} = 6.0\,V$.

$R_T = R_1 + R_2 = 1000 + 5000 = 6000\,\Omega$

$V_{supply} = I_{circuit} \times R_T$

$6 = I_{circuit} \times 6000$

$I_{circuit} = \dfrac{6}{6000} = 0.001\,A$

$V_1 = I_{circuit} \times R_1 = 0.001 \times 1000 = 1.0\,V$

$V_2 = I_{circuit} \times R_2 = 0.001 \times 5000 = 5.0\,V$

$V_{supply} = V_1 + V_2 = 6\,V$

When the resistance of R_2 increases, the voltage across R_2 (V_2) increases. The means that the voltage across R_1 (V_1) decreases, although the resistance of R_1 did not change ($V_{supply} = V_1 + V_2 = 6.0\,V$).

In any voltage divider circuit, if the resistance of one of the resistors increases then the voltage across that resistor increases and the voltage across the other resistor decreases although its resistance does not change.

For the resistors shown in Figure 3.21:

$$V_1 = I_{circuit} \times R_1 \text{ and } V_2 = I_{circuit} \times R_2$$

Therefore $\dfrac{V_1}{V_2} = \dfrac{I_{circuit} \times R_1}{I_{circuit} \times R_2}$

i.e. $\dfrac{V_1}{V_2} = \dfrac{R_1}{R_2}$

This equation can be used to calculate the voltages and resistances in a voltage divider circuit.

In the following voltage divider circuits we will be interested in what happens to the voltage across the lower component in each circuit (V_2 in Figure 3.21). Later in this chapter the voltage across the lower component will be used to switch off or switch on an LED or any other output device.

Voltage divider with a thermistor

A graph of resistance against temperature for a thermistor is given in Figure 3.22.

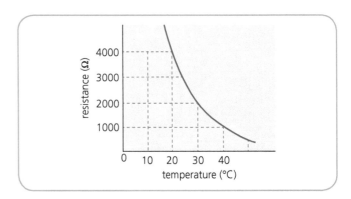

Figure 3.22 Graph of resistance against temperature for a thermistor

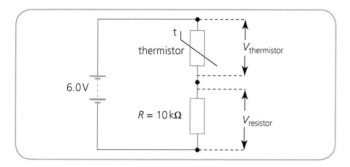

Figure 3.23 Voltage divider circuit for the results shown in Table 3.1

Figure 3.23 shows the thermistor connected in series with a 10 kΩ resistor to form a voltage divider circuit. The voltages across the thermistor and across the resistor at different temperatures are shown in Table 3.1. (You should check these values by calculation yourself and that for each row $V_{supply} = V_{thermistor} + V_{resistor} = 6V$.)

Temperature of thermistor (°C)	$V_{thermistor}$ (V)	$V_{resistor}$ (V)
20	1.7	4.3
30	1.0	5.0
40	0.5	5.5

Table 3.1

For this circuit as the temperature increases, the resistance of the thermistor decreases so the voltage across the thermistor decreases. Therefore the voltage across the resistor increases (even though the value of the resistor has not been changed).

The circuit in Figure 3.24 shows the same thermistor and resistor connected in series but with their positions interchanged. The voltages across the resistor and across the thermistor at different temperatures are shown in Table 3.2.

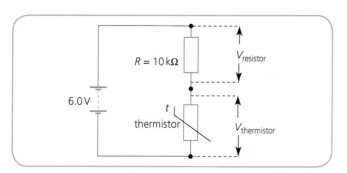

Figure 3.24 Alternative voltage divider circuit; the results for this circuit are shown in Table 3.2

Temperature of thermistor (°C)	$V_{resistor}$ (V)	$V_{thermistor}$ (V)
20	4.3	1.7
30	5.0	1.0
40	5.5	0.5

Table 3.2

For this circuit as the temperature increases, the resistance of the thermistor decreases, so the voltage across the thermistor decreases (and therefore the voltage across the resistor increases).

It should be noted that these two circuits, although containing the same components, give a voltage across the lower component that changes in opposite directions; when the temperature increases in the first circuit the voltage across the lower component increases, while in the second circuit the voltage across the lower component decreases.

Voltage divider with a light-dependent resistor

A graph of resistance for a light-dependent resistor (LDR) against light intensity is given in Figure 3.25.

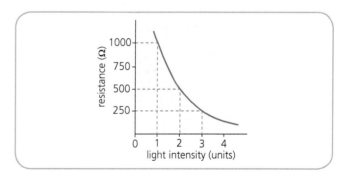

Figure 3.25 Graph of resistance against light intensity for an LDR

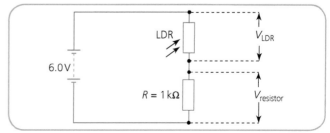

Figure 3.26 Voltage divider circuit for the results shown in Table 3.3

Figure 3.26 shows the LDR connected in series with a 1 kΩ resistor to form a voltage divider circuit. The voltages across the LDR and across the resistor at different light levels are shown in Table 3.3. (You should check these values by calculation yourself and that for each row $V_{supply} = VLDR + V_{resistor} = 6V$.)

Light intensity (units)	V_{LDR} (V)	$V_{resistor}$ (V)
1	3.0	3.0
2	2.0	4.0
3	1.2	4.8

Table 3.3

For this circuit as the light intensity increases, the resistance of the LDR decreases, so the voltage across the LDR decreases. Therefore the voltage across the resistor increases (even although the value of resistor has not been changed).

The circuit in Figure 3.27 shows the same LDR and resistor connected in series but with their positions interchanged. The voltages across the resistor and across the LDR at different light levels are shown in Table 3.4.

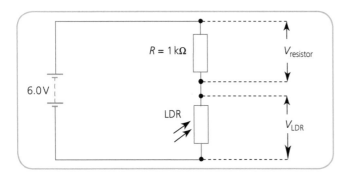

Figure 3.27 Alternative voltage divider circuit; the results for this circuit are shown in Table 3.4

Light intensity (units)	$V_{resistor}$ (V)	V_{LDR} (V)
1	3.0	3.0
2	4.0	2.0
3	4.8	1.2

Table 3.4

For this circuit as the light intensity increases, the resistance of the LDR decreases, so the voltage across the LDR decreases (and therefore the voltage across the resistor increases).

Voltage divider with a capacitor

The circuit shown in Figure 3.28 was used to time how long it took a capacitor to charge up to a certain voltage. The results for different values of capacitor and resistor are shown in Table 3.5.

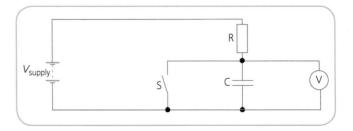

Figure 3.28 Voltage divider circuit for a capacitor and a resistor

The capacitor was discharged by closing switch S. As soon as the switch was opened, the capacitor began to charge up.

Value of C (µF)	Value of R (kΩ)	Time taken for C to charge up (s)
1000	10	60
1000	1	6
200	10	12

Table 3.5

For the circuit above, the voltage across the capacitor increases to the supply voltage, but the time taken is dependent on the values of C and R:

- Increasing the value of the capacitor increases the time taken to charge it up, i.e. a larger capacitance is able to store more charge. Capacitance ↑ = time taken ↑
- Increasing the value of the series resistor decreases the charging current and so fewer charges flow onto the capacitor plates in one second. The capacitor therefore takes longer to become fully charged. Resistance ↑ = time taken ↑

Capacitor charges up = voltage across capacitor increases (to the supply voltage).

Capacitor discharges = voltage across capacitor decreases (to zero).

Voltage divider with a switch

Figures 3.29a and b show a switch connected in series with a resistor to form a voltage divider circuit.

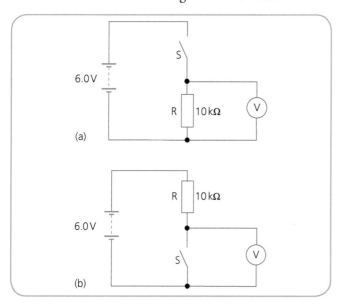

Figure 3.29 Voltage divider circuits for a switch and a resistor

In circuit (a), when the switch S is open $V = 0\,\text{V}$, and when S is closed $V = 6.0\,\text{V}$.

In circuit (b), when the switch S is open $V = 6.0\,\text{V}$, and when S is closed $V = 0\,\text{V}$.

The easiest way to understand these two circuits is to consider the switch being open, i.e. when there is no current. When the current in a resistor is zero then the voltage across the resistor is zero. Since the switch and the resistor form a voltage divider circuit and $V_{supply} = V_{resistor} + V_{switch}$, then the voltage across the switch must equal the supply voltage, i.e. 6.0 V.

Transistors

- A MOSFET (metal oxide semiconductor field effect transistor) has three terminals called the **gate**, the **source** and the **drain**. The symbol for an n-channel MOSFET is shown in Figure 3.30.
- An NPN transistor has three terminals called the **base**, the **emitter** and the **collector**. The symbol for an NPN transistor is shown in Figure 3.31.

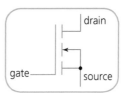

Figure 3.30 MOSFET symbol

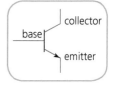

Figure 3.31 NPN transistor symbol

The MOSFET and the NPN transistor are **electronic switches** with no moving parts.

The switching of these transistors is controlled by the value of the voltage (p.d.) applied to the gate-source or base-emitter. The transistor is off (non-conducting) when the gate-source or base-emitter voltage (p.d.) is below a certain value, i.e. the electronic switch is open. However, the transistor is on (conducting) when the gate-source or base-emitter voltage (p.d.) is equal to or above this certain value, i.e. the electronic switch is closed. The value at which a transistor switches on depends on the transistor type, for instance in schools a common type of MOSFET used is fully switched on at 4.0 V while a common type of NPN transistor is fully switched on at 0.7 V.

In the following transistor circuits the gate-source or base-emitter voltage is dependent on the voltage across the lower component of the voltage divider circuit.

A temperature-controlled circuit

Figures 3.32 and 3.33 show two temperature-controlled circuits. In each circuit, the variable resistor is adjusted until at room temperature the LED is just off (the MOSFET gate-source voltage is just below a certain value and the MOSFET is non-conducting). The variable resistor now has a fixed resistance until it is again adjusted.

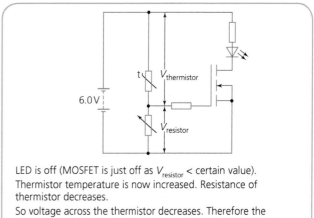

LED is off (MOSFET is just off as $V_{resistor}$ < certain value). Thermistor temperature is now increased. Resistance of thermistor decreases.
So voltage across the thermistor decreases. Therefore the voltage across the (variable) resistor increases and when it is ≥ certain value, MOSFET switches on (conducting) and LED lights.

Figure 3.32 A temperature-controlled circuit

31

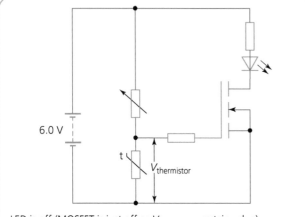

LED is off (MOSFET is just off as $V_{thermistor}$ < certain value).
Thermistor is now cooled. Resistance of thermistor increases.
So voltage across the thermistor increases and when it is ≥ certain value, MOSFET switches on and LED lights.

Figure 3.33 An alternative temperature-controlled circuit

Note: A variable resistor is used in this type of circuit instead of a fixed value of resistor. The variable resistor allows the circuit to be adjusted to different input conditions (temperature in this case) before the LED (output device) comes on.

A light-controlled circuit

Figures 3.34 and 3.35 show two light-controlled circuits. The variable resistor is adjusted, in each circuit, until at normal light level the LED is just off (the NPN transistor base-emitter voltage is just below a certain value and the transistor is non-conducting).

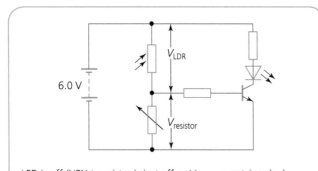

LED is off (NPN transistor is just off as $V_{resistor}$ < certain value).
More light is now shone on LDR. Resistance of LDR decreases. Voltage across the LDR decreases. Therefore voltage across the (variable) resistor increases and when it is ≥ certain value, transistor switches on and LED lights.

Figure 3.34 A light-controlled circuit

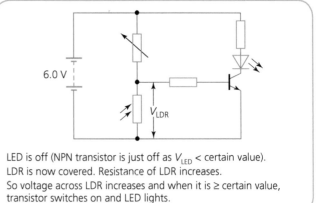

LED is off (NPN transistor is just off as V_{LED} < certain value).
LDR is now covered. Resistance of LDR increases.
So voltage across LDR increases and when it is ≥ certain value, transistor switches on and LED lights.

Figure 3.35 An alternative light-controlled circuit

A time-controlled circuit

Figures 3.36 and 3.37 show two time-controlled circuits. The switch S is closed to discharge the capacitor (voltage across capacitor = 0 V), then opened to allow the capacitor to start charging.

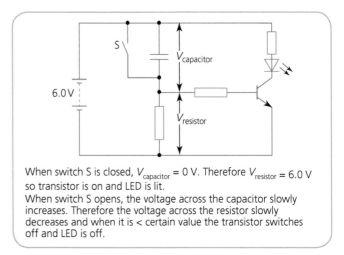

When switch S is closed, $V_{capacitor}$ = 0 V. Therefore $V_{resistor}$ = 6.0 V so transistor is on and LED is lit.
When switch S opens, the voltage across the capacitor slowly increases. Therefore the voltage across the resistor slowly decreases and when it is < certain value the transistor switches off and LED is off.

Figure 3.36 Time-controlled circuit

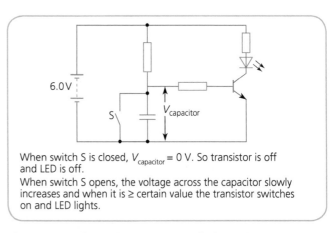

When switch S is closed, $V_{capacitor}$ = 0 V. So transistor is off and LED is off.
When switch S opens, the voltage across the capacitor slowly increases and when it is ≥ certain value the transistor switches on and LED lights.

Figure 3.37 Alternative time-controlled circuit

In both cases the time taken for the LED to go off or come on is dependent on the values of the capacitor and the resistor.

A switch-controlled circuit

Figures 3.38 and 3.39 show two switch-operated circuits.

When a switch is open in either circuit, there is no current in the resistor and so the voltage across the resistor is zero and therefore the voltage across the switch is 6.0 V.

With switch S open, V_{switch} = 6.0 V. So transistor is on and LED lights.
With switch S closed, V_{switch} = 0 V. So transistor switches off and LED is off.

Figure 3.39 Alternative switch-controlled circuit

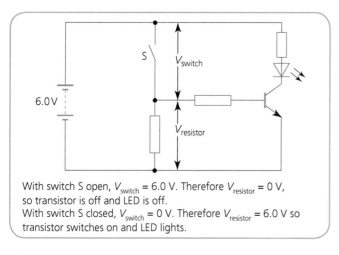

With switch S open, V_{switch} = 6.0 V. Therefore $V_{resistor}$ = 0 V, so transistor is off and LED is off.
With switch S closed, V_{switch} = 0 V. Therefore $V_{resistor}$ = 6.0 V so transistor switches on and LED lights.

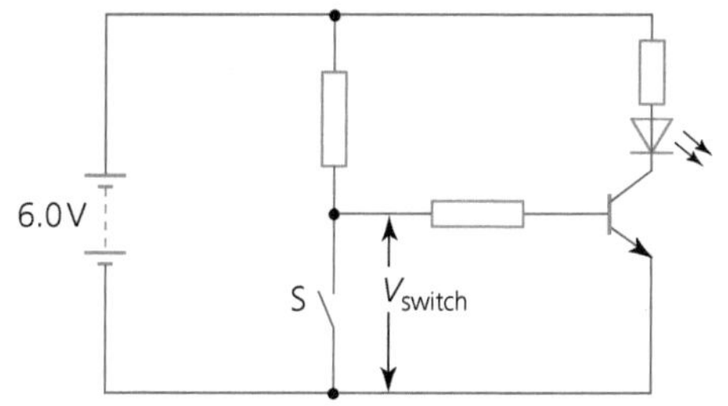

Figure 3.38 Switch-controlled circuit

Key facts and physics equations: electrical components and electronic circuits

- Some devices change a form of energy into electrical energy, e.g. microphones, solar cells and thermocouples.
- Some devices change electrical energy into another form of energy, e.g. loudspeakers, electric motors, filament lamps and light-emitting diodes (LEDs).
- The resistance of a thermistor changes with temperature – the resistance of most thermistors decreases as the temperature increases.
- The resistance of an LDR decreases with increasing light level.

- LEDs will only light up when connected to a d.c. supply the correct way round.
- A resistor should always be connected in series with an LED – this protects the LED from damage from too high a current in the LED (or too high a voltage across the LED).
- A diode will only conduct when connected to a d.c. supply the correct way round.
- A MOSFET and an NPN transistor are electrically operated switches.
- A transistor is non-conducting (off) for voltages below a certain value but conducting (on) at voltages at or above this certain value.

Electricity

End-of-chapter questions

1 Identify the circuit symbols shown below.

(a) —(M)—

(b)

(c)

(d) ⊳⊢

(e)

(f)

(g)

(h)

(i)

(j)

(k)

(l)

(m)

Figure 3.40

2 State the energy conversion for the following:
  a) a microphone
  b) a solar cell
  c) a loudspeaker
  d) an electric motor
  e) a filament lamp
  f) an LED.

3 A circuit is set up as shown in Figure 3.41.

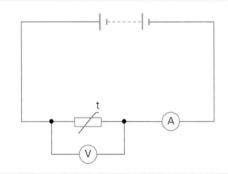

Figure 3.41

The temperature of the thermistor is 18 °C. The reading on the voltmeter is 6.0 V. The reading on the ammeter is 12 mA.
  a) Calculate the resistance of the thermistor at this temperature.
  b) The temperature of the thermistor rises to 20 °C. The voltmeter reading remains at 6.0 V. Suggest a value for the reading on the ammeter.

4 A student designs a circuit to light an LED. The student uses the following components: a 6.0 V battery, a switch, an LED and a resistor.
  a) Draw a suitable diagram that will allow the LED to light safely when the switch is closed.
  b) The maximum voltage across the LED must not exceed 1.75 V. The maximum current in the LED must not exceed 11 mA. Calculate the value of the resistor required for the circuit.

5 In the following circuits calculate the voltages V_1 and V_2.

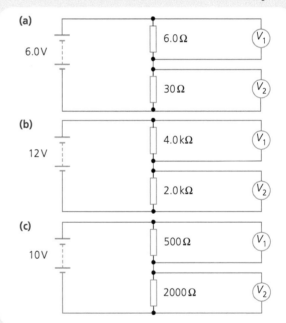

Figure 3.42

6 A circuit is set up as shown.

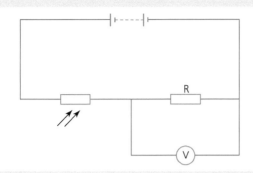

Figure 3.43

The light intensity falling on the LDR increases. Describe how the reading on the voltmeter changes when the light intensity increases.

7 A student sets up the circuit shown below.

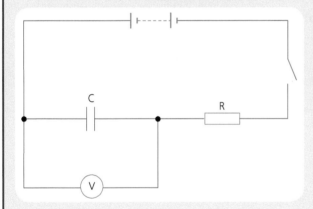

Figure 3.44

When the switch is open the reading on the voltmeter is zero. The switch is now closed.

a) What happens to the reading on the voltmeter as the capacitor charges?

b) The capacitor takes 30 s to charge to 5.0 V. Suggest **one** change to the circuit that would give a longer time of charge.

8 The diagram shows an electronic circuit.

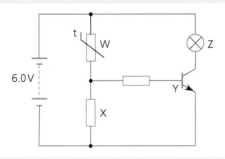

Figure 3.45

a) Name the components W, X, Y and Z in the circuit.

b) What is the purpose of component Y in the circuit?

9 A student builds a circuit to detect when the element of an electric cooker is hot (Figure 3.46).

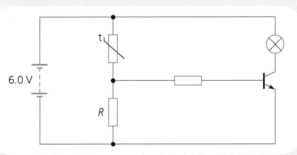

Figure 3.46

The resistance of the thermistor decreases as its temperature increases. Explain what will happen as the temperature of the cooker element increases.

10 The diagram below shows a circuit to detect the light level in a room.

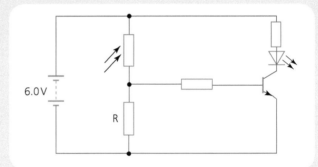

Figure 3.47

The room is initially dark. Describe and explain what will happen as the light level in the room increases.

11 A student builds a circuit so that a motor will only rotate in one direction. The student uses a 6.0 V battery, a diode and a motor.

Draw a suitable diagram that will allow the motor to rotate in only one direction.

12 Part of the circuit that operates an automatic light is shown below.

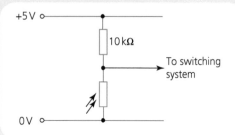

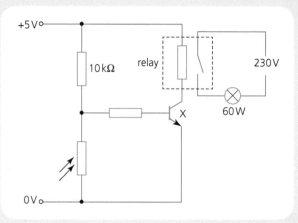

Figure 3.48

The resistance of the light-dependent resistor (LDR) in different lighting conditions is shown in the table below.

Lighting condition	Resistance of LDR (kΩ)
Light	0.5
Dark	90

a) The LDR is placed in darkness. Calculate the voltage across the LDR.

b) A lamp rated at 60 W, 230 V is connected to the light sensor using the circuit shown in Figure 3.49.

Figure 3.49

i) Name component X.

ii) When the LDR is in darkness the 60 W lamp lights. Explain how this happens.

iii) Calculate the current in the 60 W lamp when it is lit.

Properties of Matter

4 Heat

At the end of this chapter you should be able to:

1 Use the following terms correctly in context: temperature; heat; Celsius.
2 State that the temperature of a substance is a measure of the average kinetic energy of the particles of the substance.
3 State that heat can be transferred from a higher temperature to a lower temperature by the processes of conduction, convection and radiation.
4 State that the heat loss every second from a hot object depends on the temperature difference between the object and its surroundings.
5 State that the same mass of different materials requires different quantities of energy to change their temperature by 1 degree Celsius.
6 Carry out calculations involving energy, mass, specific heat capacity and temperature change.
7 State that energy is gained or lost by a substance when its state is changed.
8 State that a change of state does not involve a change in temperature.
9 Carry out calculations involving energy, mass and specific latent heat.
10 Use the following terms correctly in context: specific heat capacity; change of state; latent heat of fusion; latent heat of vaporisation.
11 Carry out calculations involving energy, work, power and the principle of conservation of energy.

Heat and temperature

Heat, just like light and sound, is a form of energy. It is measured in joules (J). Temperature is a measure of how hot or cold a substance is and is measured in degrees Celsius (°C).

The temperature of a substance is a measure of the average kinetic energy of all the particles of the substance. The higher the temperature of a substance, the greater the average kinetic energy of the particles of the substance.

Heat transfer

Heat is always transferred from a higher temperature to a lower temperature. There are three possible ways in which heat can be transferred, called **conduction**, **convection** and **radiation**.

Conduction

Heat can be transferred through materials. Heat moves from the high temperature to the low temperature.

Materials that allow heat to move easily through them are called conductors – metals are the best conductors due to free electron heat transfer (see below). Materials that do not allow heat to move through them easily are called insulators – non-metal solids, liquids and gases are good insulators (poor conductors).

The electrons in metals can leave their atoms and move about in the metal as free electrons – this is why metals are good conductors of electricity. The parts of the metal atoms left behind are now charged metal ions. As the metal is a solid, the ions are packed closely together and they are continually vibrating. The hotter the metal, the greater the vibrations and the more kinetic energy the ions have. This kinetic energy is transferred from hot parts to cooler parts of the metal by the free electrons. The electrons move through the structure of the metal, colliding with ions as they go and so transferring energy.

Convection

Heat can be transferred by the movement of the heated particles making up a liquid or gas. The heated fluid (liquid or gas) becomes less dense and rises, cold fluid falls to take its place, and convection currents are set up. Convection cannot take place in a solid because the particles cannot move away.

Many heat insulators contain trapped air – for example, cotton wool, felt and woollen clothes. No convection can take place because the air is trapped and cannot move. Air is also a non-metal, so it is a poor conductor of heat.

Radiation

Radiation travels in straight lines until absorbed by an object. It can travel through a vacuum (the Earth is heated by radiation from the Sun). Heat radiation travels at a speed of $3 \times 10^8 \, \text{m s}^{-1}$. All hot materials radiate heat. Heat radiation is also known as infrared radiation.

Physics beyond the classroom

Scientists are using telescopes, satellites and computers to study the interior of the Sun. They have discovered that the Sun has a number of different layers.

The Sun has a central core. All the solar energy is generated in the core by nuclear fusion (see Chapter 8). Around the core there is a radiative zone. In this region, energy is transported by radiation. Above the radiative zone there is the convection zone. Energy is transported by convection in this region. The surface of the convection zone is where light is created.

Physics beyond the classroom

All houses lose heat by conduction, convection and radiation to the colder outside. To have a comfortable living temperature in the house but avoid high fuel bills it is important to reduce these heat losses. The main heat losses from a house and their prevention are shown in Figure 4.1.

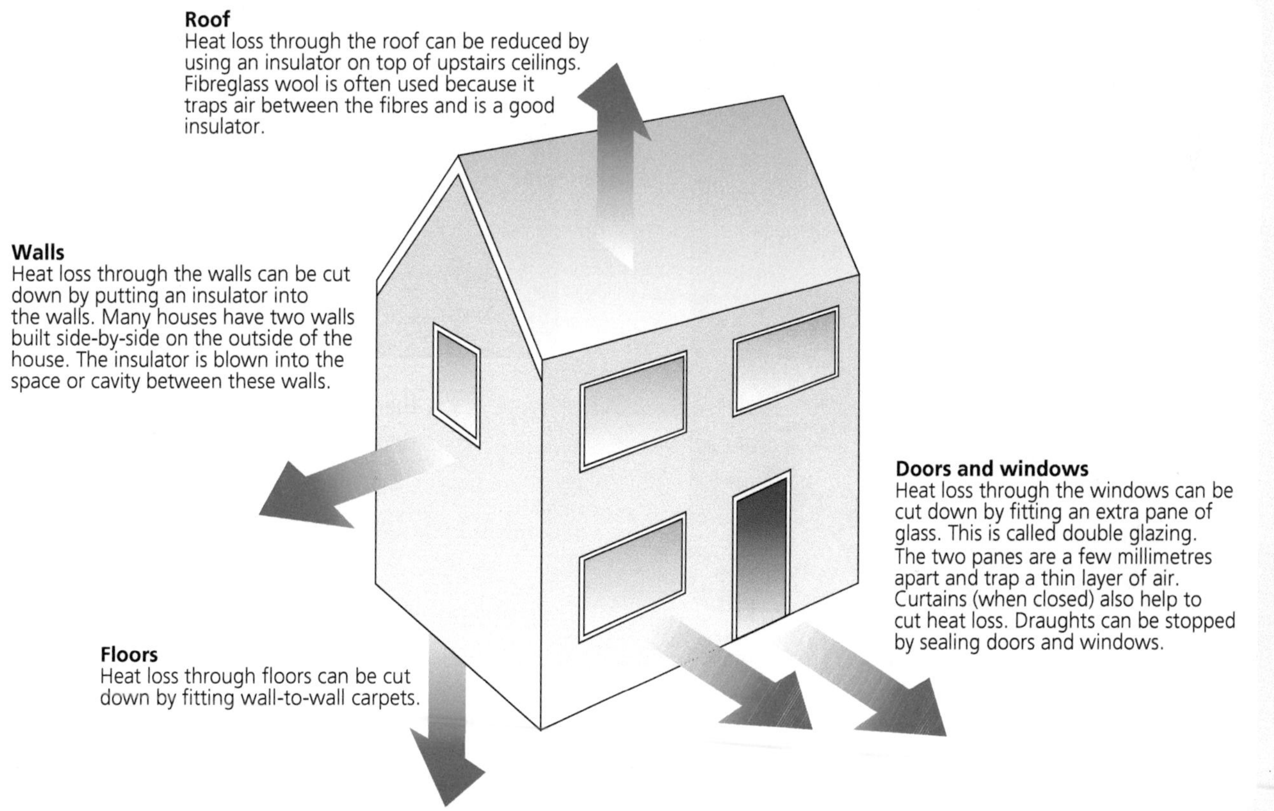

Roof
Heat loss through the roof can be reduced by using an insulator on top of upstairs ceilings. Fibreglass wool is often used because it traps air between the fibres and is a good insulator.

Walls
Heat loss through the walls can be cut down by putting an insulator into the walls. Many houses have two walls built side-by-side on the outside of the house. The insulator is blown into the space or cavity between these walls.

Doors and windows
Heat loss through the windows can be cut down by fitting an extra pane of glass. This is called double glazing. The two panes are a few millimetres apart and trap a thin layer of air. Curtains (when closed) also help to cut heat loss. Draughts can be stopped by sealing doors and windows.

Floors
Heat loss through floors can be cut down by fitting wall-to-wall carpets.

Figure 4.1 Main heat losses from a house

Physics beyond the classroom

Warm objects give off invisible 'heat rays' called infrared radiation. These rays are invisible to our eyes but can be viewed using an infrared camera as shown in Figure 4.2.

In some cameras, colour images showing different temperatures are produced. These pictures are called thermograms and can be particularly useful in medicine. For example, Figure 4.3 shows the thermogram of an arthritic elbow of a patient.

The orange/yellow colours on this thermogram show abnormally high temperatures indicating inflammation of the elbow joint.

Physiotherapists also use infrared radiation to penetrate the skin and to heat damaged muscles. Heat causes the muscles to heal more rapidly.

Figure 4.2 (Top) picture taken with a 'normal' camera and (bottom) the same picture taken with an infrared camera

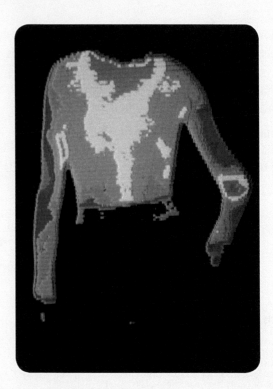

Figure 4.3 A thermogram

During re-entry to the Earth's atmosphere any space vehicle heats up due to the frictional effects of the atmosphere on the very fast-moving object.

The space shuttle used thousands of thermal silicon tiles (Figure 4.4) to protect it from the extremely high temperatures produced during re-entry. The black tiles mainly covered the lower parts of the shuttle, which experienced the highest temperatures (up to 1200 °C). The white tiles covered the rest of the shuttle, which experienced lower temperatures (up to 600 °C).

The space shuttle is no longer used to transport people and materials into space, having been 'retired' from service in 2011.

Figure 4.4

Heat loss and temperature difference

David and Mary decided to investigate how heat loss depends on the temperature of the surroundings. They heated two identical metal blocks to 100 °C. One block was placed in cold water and the other in warm water. The temperature of each block was measured every minute for 10 minutes. The results were displayed on a graph, as shown in Figure 4.5.

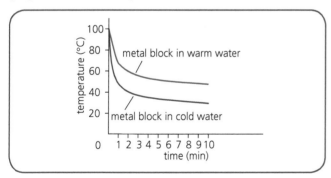

Figure 4.5 Graphs of temperature against time for both metal blocks

From their investigation they concluded that the metal block in cold water lost energy more quickly than the block in warm water.

The amount of energy lost every second depends on the difference in temperature between an object and its surroundings:

- The larger the difference in temperature, the larger the loss of energy in 1 second.

- The smaller the difference in temperature, the smaller the loss of energy in 1 second.

Changing temperature

When you want to make a hot drink you put some water in the kettle and switch the kettle on. The water gets hot as a result of the energy supplied by the heating element of the kettle. However, the number of joules of energy required to warm the water depends on:

- the temperature rise – more energy is required for a larger temperature rise
- the mass of water – more energy is required for a larger mass of water.

Example

The minimum energy required to raise the temperature of 0.5 kg of water by 10 °C is 20 900 J.

a) What is the minimum energy required to raise the temperature of 0.5 kg of water by 20 °C?
b) What is the minimum energy required to raise the temperature of 1.0 kg of water by 40 °C?

Solution

a) It requires 20 900 J to change the temperature of 0.5 kg of water by 10 °C.
   It requires twice as much energy to change the temperature of 0.5 kg by 20 °C (2 × 10 °C).
   It requires 41 800 J to change the temperature of 0.5 kg of water by 20 °C.

b) It requires 20 900 J to change the temperature of 0.5 kg of water by 10 °C.
It requires (2 × 20 900) J to change the temperature of (2 × 0.5) kg of water by 10 °C.
It requires 41 800 J to change the temperature of 1.0 kg of water by 10 °C.
It requires (4 × 41 800) J to change the temperature of 1.0 kg of water by (4 × 10 °C).
It requires 167 200 J to change the temperature of 1.0 kg of water by 40 °C.

Specific heat capacity

So far we have only considered heating the same material, namely water. However, if equal masses of water and copper are supplied with the same number of joules of heat then you would find that the rise in temperature of the copper would be higher.

The energy (E_h) required to change the temperature of a material depends on:

- the change in temperature of the material (ΔT)
- the mass of the material (m)
- a constant for the material – the **specific heat capacity** of the material, c.

The specific heat capacity of a material is the number of joules of energy required to change the temperature of 1.0 kg of that material by 1.0 °C. The unit of specific heat capacity is joules per kilogram per degree Celsius ($J\,kg^{-1}\,°C^{-1}$).

Water has a specific heat capacity of 4180 joules per kilogram per degree Celsius ($4180\,J\,kg^{-1}\,°C^{-1}$). This means that **4180 joules of energy are required to change the temperature of 1.0 kg of water by 1.0 °C**. It follows that:

- It requires 16 720 J to change the temperature of 4.0 kg of water by 1.0 °C ($4 × E_h$, as $4 × m$).
- It requires 83 600 J to change the temperature of 4.0 kg of water by 5.0 °C ($5 × E_h$, as $5 × \Delta T$).

The energy needed to change the temperature of a material can be represented by the equation:

$$E_h = cm\Delta T$$

where E_h = energy needed to change the temperature of the material, measured in joules (J),

c = specific heat capacity of the material, measured in joules per kilogram per degree Celsius ($J\,kg^{-1}\,°C^{-1}$),
m = mass of the material, measured in kilograms (kg),
ΔT = change in temperature of the material, measured in degrees Celsius (°C).

A table of specific heat capacities for selected materials is given on the *Data Sheet* on page 170.

Worked example

Example

The mass of an aluminium block is 750 g. The temperature of the block falls from 65 °C to 18 °C. Calculate the energy released by the block.

Solution

Note: c for aluminium from *Data Sheet*
Note: 750 g = 750 × 10⁻³ kg = 0.750 kg

$E_h = cm\Delta T$

$E_h = 902 × 750 × 10^{-3} × (65 - 18)$

$E_h = 902 × 750 × 10^{-3} × 47$

$E_h = 3.18 × 10^4\,J$

Heat problems

Energy can be changed from one form to another. However, the total amount of energy remains unchanged – this is the **principle of conservation of energy**.

An electric heater converts electrical energy into an equal amount of heat. Due to conduction, convection and radiation, some of the heat supplied by the heater will be transferred ('lost') to the surroundings. This means that the material that is being heated will absorb (take in) less energy than was supplied by the heater:

$$\text{energy supplied} = \text{energy absorbed by material} + \text{energy transferred to the surroundings}$$

However, some heat problems use the term 'well-insulated container' – this means that the container does not transfer or 'lose' energy to the surroundings and that the container itself does not absorb energy from the contents. In these cases:

$$\text{energy supplied} = \text{energy absorbed by material}$$

Worked examples

Example 1

A well-insulated container contains 0.5 kg of water. The initial temperature of the water is 16 °C. A heater is immersed in the water and switched on. The heater supplies 83 600 J of energy to the water. Calculate the final temperature of the water.

Solution

Note: c for water from *Data Sheet*
Well insulated means that no energy is transferred to the surroundings.

energy supplied by heater = energy absorbed by water

$$E_h = cm\Delta T$$
$$83\,600 = 4180 \times 0.5 \times \Delta T$$
$$83\,600 = 2090 \times \Delta T$$
$$\Delta T = \frac{83\,600}{2090} = 40\,°C$$

final temperature = initial temperature + ΔT
$$= 16 + 40 = 56\,°C$$

Example 2

A well-insulated kettle contains 0.8 kg of water. The kettle is switched on for 2 minutes. The temperature of the water rises from 16 °C to 100 °C. Calculate the power rating of the element of the kettle.

Solution

Note: 2 minutes = (2 × 60) s
Note: c for water from *Data Sheet*
Assuming no energy is transferred to the surroundings then:

energy supplied by kettle = energy absorbed by water

$$E_h = cm\Delta T = 4180 \times 0.8 \times (100 - 16)$$

$$E_h = 4180 \times 0.8 \times 84 = 280\,896\,J$$

$$P = \frac{E}{t} = \frac{280\,896}{2 \times 60} = 2341\,W = 2.3\,kW$$

Example 3

A deep-fat fryer contains cooking oil. The mass of the cooking oil is 800 g. The deep-fat fryer is switched on for 180 s. The temperature of the cooking oil rises from 20 °C to 140 °C in this time. The specific heat capacity of the cooking oil is 3000 J kg^{-1} °C^{-1}.

a) Calculate the minimum power rating of the element of the deep-fat fryer.
b) The deep-fat fryer operates from the 230 V mains. When switched on, calculate the current in the element of the deep-fat fryer.

Solution

a) Note: 800 g = 800 × 10^{-3} kg = 0.8 kg
   Assuming no energy is transferred to the surroundings then:

$$\frac{\text{energy supplied}}{\text{by element}} = \frac{\text{energy absorbed}}{\text{by cooking oil}}$$

$$E_h = cm\Delta T$$
$$E_h = 3000 \times 800 \times 10^{-3} \times (140 - 20)$$
$$E_h = 3000 \times 800 \times 10^{-3} \times 120 = 288\,000\,J$$
$$P \times t = 288\,000$$
$$P = \frac{E}{t} = \frac{288\,000}{180} = 1600\,W = 1.6\,kW$$

b) $$P = IV$$
$$1600 = I \times 230$$
$$I = \frac{1600}{230} = 6.96\,A$$

Specific latent heat

When cold water in a kettle is heated, its temperature rises until the water starts to boil at 100 °C. Further heating of the water no longer produces a rise in temperature of the water but steam is produced at 100 °C. The energy supplied by the kettle is now being used to change water at 100 °C into steam at 100 °C.

The energy required to change 1.0 kg of a liquid at its boiling point into 1.0 kg of vapour at the same temperature is called the **specific latent heat of vaporisation**.

The word latent means 'hidden' and refers to the fact that the temperature of the material does not change and the energy supplied to the material seems to have 'disappeared'.

When ice at its melting point of 0 °C is heated it turns into water at 0 °C. Energy is required to change the ice into water without a change in temperature.

The energy required to change 1.0 kg of a solid at its melting point into 1.0 kg of liquid at the same temperature is called the **specific latent heat of fusion**.

The three states of matter are solid, liquid and gas. Whenever a material changes state, latent heat is required. When a material changes from a solid to a liquid or a liquid to a gas, energy is needed to break down the forces (or bonds) holding the particles together and to push the particles further apart.

● Specific latent heat of fusion is the energy required to change 1.0 kg of a solid at its melting point into a liquid without change in temperature.
● Specific latent heat of vaporisation is the energy required to change 1.0 kg of a liquid at its boiling point into a gas without change in temperature.

When a material changes state from solid to liquid or liquid to gas, latent heat is absorbed (taken in). When a material changes state from gas to liquid or liquid to solid, latent heat is released (given out). When a material changes state there is no change in temperature.

The symbol for specific latent heat is l. Specific latent heat is measured in $J\,kg^{-1}$.

The specific latent heat of a material is the energy required to change the state of 1 kg of the material without a change in temperature.

For a material with a specific latent heat l, the energy (E_h) required is as follows:

● To change the state of 1.0 kg of the material at constant temperature requires l J.
● To change the state of m kg of the material at constant temperature requires $m \times l$ J.

So,

$$E_h = ml$$

where E_h = energy needed to change state of material, measured in joules (J),
m = mass of material that changed state, measured in kilograms (kg),
l = specific latent heat of material, measured in joules per kilogram ($J\,kg^{-1}$).

Tables of specific latent heat of fusion and specific latent heat of vaporisation for selected materials are given on the *Data Sheet* on page 170.

A sample of a solid is heated. The graph in Figure 4.6 shows how the temperature of the sample changes with the energy absorbed by it.

Between A and B the sample is in the solid state – the energy required to change its temperature in the solid state is given by $E_{h\ solid} = cm\Delta T_{solid}$ where c is the specific heat capacity of the solid.

Between B and C the sample is changing state (from solid to liquid) – the energy required to change its state is given by $E_{h\ fusion} = ml$ where l is the specific latent heat of fusion of the sample.

Between C and D the sample is in the liquid state – the energy required to change its temperature in the liquid state is given by $E_{h\ liquid} = cm\Delta T_{liquid}$ where c is the specific heat capacity of the liquid.

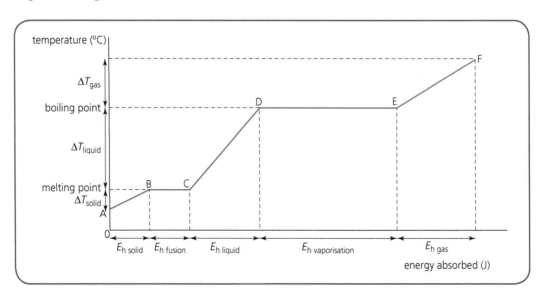

Figure 4.6

Between D and E the sample is changing state (from liquid to gas) – the energy required to change the state is given by $E_{h\,vaporisation} = ml$ where l is the specific latent heat of vaporisation of the sample.

Between E and F the sample is in the gas state – the energy required to change its temperature in the gas state is given by $E_{h\,gas} = cm\Delta T_{gas}$ where c is the specific heat capacity of the gas.

Worked examples

Example 1

Ice cubes at 0 °C are placed in a well-insulated container. The mass of the ice cubes is 0.6 kg. A heater is now placed in the ice in the container. The power rating of the heater is 50 W. The heater is switched on. Calculate the time taken for the heater to melt 0.04 kg of the ice. The mass of the ice melted by heat from the room can be neglected.

Solution

Note: l_{fusion} for water from *Data Sheet*

Note: m is the mass of the material that changes state

energy required to melt ice $E_h = ml$

$E_h = 0.04 \times 3.34 \times 10^5 = 13\,360\,J$

Assuming no energy is transferred to the surroundings then:

$$\begin{array}{l}\text{energy supplied} \\ \text{by heater}\end{array} = \begin{array}{l}\text{energy absorbed by ice} \\ \text{changing state}\end{array}$$

$P \times t = 13\,360$

$50 \times t = 13\,360$

$t = \dfrac{13\,360}{50} = 267\,s$

Example 2

A teacher uses the apparatus shown in Figure 4.7 to obtain a value for the specific latent heat of vaporisation of water.

The lid of the kettle is left off so that the kettle does not switch off. When the water is boiling the following information is recorded:

- mass of water changed to steam = 100 g
- power rating of heater = 2200 W
- time taken to change 100 g of water into steam = 105 s

a) Calculate the specific latent heat of vaporisation of water obtained from this experiment.

b) Explain why this value is likely to be higher than the accepted value for the specific latent heat of vaporisation of water.

Solution

a) Energy supplied by heater:

$E_h = P \times t = 2200 \times 105 = 231\,000\,J$

Note: $100\,g = 100 \times 10^{-3}\,kg = 0.100\,kg$

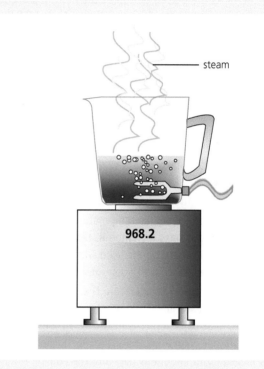

steam

968.2

Figure 4.7 An experiment to obtain a value for the specific latent heat of vaporisation of water

Assuming no energy is transferred to the surroundings then:

$$\begin{array}{l}\text{energy supplied} \\ \text{by heater}\end{array} = \begin{array}{l}\text{energy absorbed by} \\ \text{water changing state}\end{array}$$

$E_h = ml$

$231\,000 = 100 \times 10^{-3} \times l$

$l = \dfrac{231\,000}{100 \times 10^{-3}} = 2.31 \times 10^6\,J\,kg^{-1}$

b) Energy supplied by heater will be larger than the energy absorbed by water changing state, as energy will be lost to the surroundings.

(energy supplied = energy required to change the state of 0.1 kg of water + energy lost to surroundings)

Since $l = \dfrac{E_h}{m}$ and E_h is larger than it should be to change the state of 0.1 kg of water, then l will be higher than the accepted value.

Physics beyond the classroom

Latent heat is used to help keep food in a 'cool box', the type used for picnics, cool.

Special chemical packs are placed in a freezer. The chemicals are a liquid at normal room temperature but change to a solid during cooling in a freezer. The frozen chemical packs are placed on top of the food in the cool box – air surrounding the frozen packs cools and falls due to convection. Initially the cold frozen packs keep the box cool, as they take heat from the food.

However, the temperature in the cool box will rise slowly and the chemicals will begin to melt, i.e. change from a solid to a liquid. The energy needed to bring about this change in state is absorbed from the food. The food 'loses' heat and so is kept cool for a longer period.

Key facts and physics equations: heat

- Temperature is a measure of the average kinetic energy of the particles making up a substance.
- Equal masses of different materials require different amounts of energy to change their temperature by 1 °C.
- Energy needed to change temperature = specific heat capacity × mass × change in temperature, i.e. $E_h = cm\Delta T$. This equation is used whenever there is a change in temperature in a material.
- During a change in temperature the energy absorbed or lost by a material, E_h, is measured in joules (J), specific heat capacity, c, is measured in joules per kilogram per degree Celsius ($J\,kg^{-1}\,°C^{-1}$), mass, m, is measured in kilograms (kg) and the change in temperature, ΔT, is measured in degrees Celsius (°C).
- A specific heat capacity, c, of $100\,J\,kg^{-1}\,°C^{-1}$ means that 100 J of energy are required to change the temperature of 1.0 kg of the material by 1.0 °C.

- A change of state occurs when a solid changes into a liquid (or a liquid changes to a solid) or a liquid changes to a gas (or a gas changes to a liquid).
- There is no change in temperature when a change of state occurs.
- Energy needed to change state = mass that changes state × specific latent heat, i.e. $E_h = ml$. This equation is used whenever there is a change in state of the material.
- During a change in state, the energy absorbed or lost by the material, E_h, is measured in joules (J), mass that changes state, m, is measured in kilograms (kg) and the specific latent heat of fusion or vaporisation, l, is measured in joules per kilogram ($J\,kg^{-1}$).
- When a material changes temperature it cannot change state and when a material changes state it cannot change temperature.

End-of-chapter questions

Information, if required, for use in the following questions can be found on the *Data Sheet* on page 170.

1   The minimum energy required to raise the temperature of 2.0 kg of water by 1.0 °C is 8360 J. What is the minimum energy required to raise the temperature of:
    a) 4.0 kg of the water by 1.0 °C
    b) 4.0 kg of the water by 5.0 °C
    c) 8.0 kg of the water by 10 °C?

2   Calculate the minimum energy required to raise the temperature of:

a) a sample of water of mass 0.5 kg from 18 °C to 58 °C
b) a iron baking tray of mass 0.95 kg from 18 °C to 198 °C.

3   Water has a specific heat capacity of $4180\,J\,kg^{-1}\,°C^{-1}$. State what is meant by the term *specific heat capacity*.

4   A well-insulated kettle contains 1.2 kg of water at a temperature of 20 °C. The kettle is switched on for 180 s. The temperature of the water rises to 100 °C.
    a) Calculate the energy absorbed by the water in 180 s.
    b) Calculate the power rating of the heating element of the kettle.

5 A heater is connected to a 12 V supply. When the heater is operating, the current in the heater is 4.0 A. The heater is used to heat a 1.0 kg copper block. The heater is switched on for 5 minutes.
   a) Calculate the energy produced by the heater in 5 minutes.
   b) Calculate the maximum possible rise in the temperature of the copper block.
   c) Explain why the rise in temperature of the copper block will be less than the answer to b).

6 A kettle contains 0.4 kg of water. The initial temperature of the water is 18 °C. The element of the kettle has a power rating of 2000 W. The kettle is switched on for 40 s.
   a) Calculate the energy produced by the element of the kettle in 40 s.
   b) Calculate the maximum final temperature of the water.
   c) Why is this the maximum value for the final temperature?

7 Calculate the minimum energy required to change 0.8 kg of ice at 0 °C into water at 0 °C.

8 Calculate the minimum energy required to change 0.2 kg of water at 100 °C into steam at 100 °C.

9 State what is meant by term *specific latent heat of fusion*.

10 The power rating of a heater is 2000 W. The heater is used, in a well-insulated container, to bring a sample of water to its boiling point. After the water has reached its boiling point the heater is left on for a further 80 s. Calculate the mass of water changed to steam in this time.

11 A container is placed on a digital balance. A sample of water at 100 °C is poured into the container. A heater is now placed in the water. The power rating of the heater is 50 W. The reading on the balance is 285 g. The heater is switched on for 6 minutes. The reading on the balance falls to 278 g. Calculate the specific latent heat of vaporisation of water.

12 A sample of water of mass 100 g is poured into a well-insulated container. The temperature of the water is 80 °C. The mass of a second sample of water is 200 g. The temperature of this sample is 20 °C. The second sample is now added to the container. Calculate the minimum temperature of the mixture of water in the container.

13 In a research laboratory, an investigation is carried out to see the effect of condensed steam on a piece of tissue. The sample of steam is at a temperature of 100 °C. The mass of the sample of steam is 35 g. The steam is allowed to condense onto the piece of tissue. The final temperature of the condensed steam is 40 °C. Calculate the total energy released by the sample of steam.

5 Gas laws and the kinetic model

Learning outcomes

At the end of this chapter you should be able to:

1 State that pressure is the force per unit area, when the force acts at right angles to the surface.
2 State that 1 pascal is 1 newton per square metre.
3 Carry out calculations involving pressure, force and area.
4 Describe how the kinetic model accounts for the pressure exerted by a gas.
5 State that the pressure exerted by a fixed mass of gas at constant temperature is inversely proportional to its volume.
6 State that the pressure exerted by a fixed mass of gas at constant volume is directly proportional to its temperature measured in Kelvin (K).

7 State that the volume of a fixed mass of gas at constant pressure is directly proportional to its temperature measured in Kelvin (K).
8 Carry out calculations to convert temperatures in °C to K and vice versa.
9 Carry out calculations involving pressure, volume and temperature of a fixed mass of gas using the general gas equation.
10 Explain what is meant by absolute zero of temperature.
11 Explain the pressure–volume, pressure–temperature and volume–temperature laws qualitatively in terms of a kinetic model.

Kinetic theory of matter

In Chapter 4 we looked at the three states of matter – solid, liquid and gas. We are unable to see what makes up a solid, liquid or gas with the human eye – or indeed with a very powerful light microscope – but if we were able to 'see' matter, what would it look like?

Scientists came up with a 'theory' or 'model' of what they think matter looks like – it is known as the **kinetic theory of matter**. According to this theory the three states of matter can be thought of as being made up of particles, called atoms or molecules. These particles are in continuous motion. Acting between the particles are forces of attraction.

The solid state

In the solid state, the particles are closely packed together in a definite shape. They do not move relative to each other but are able to vibrate about a fixed

position. This explains why most solids have a definite shape and suggests that neighbouring particles are kept in position by very large forces of attraction. Heating a solid gives the particles more energy. This makes the particles vibrate more violently about their fixed position. The average kinetic energy of the particles increases, which means the temperature of the solid increases.

The liquid state

Supplying more energy causes the temperature of the solid to increase until at a certain temperature, called the **melting point**, the particles partly overcome the forces of attraction, which kept them in a fixed position, and the solid changes to a liquid. The particles are now able to slip over one another in all directions. This is why a liquid can be poured and why it takes up the shape of the container into which it is poured. Since the particles are not so tightly bound there are small spaces between them.

The gas state

Supplying more energy causes the temperature of the liquid to increase until at a certain temperature, called the **boiling point**, the particles can completely overcome the forces between them and they go off on their own – the liquid changes to a gas. In the gas state the particles are moving very fast in all directions. They are far apart (compared with their size) so that the forces of attraction between them can be neglected. This accounts for the fact that a gas will spread throughout a room. The gas particles undergo random motion.

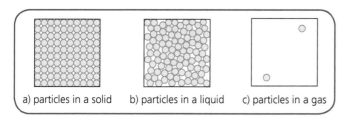

a) particles in a solid   b) particles in a liquid   c) particles in a gas

Figure 5.1 States of matter

- Solids, liquids and gases are made up of tiny particles.
- The particles of a solid are only able to vibrate about a fixed position.
- The particles of a liquid are able to slip over one another.
- The particles of a gas move around in all directions.

Pressure

The pressure acting on a surface is defined as the force acting at right angles on unit area of a surface.

$$pressure = \frac{force\ exerted}{area\ acted\ on}$$

$$p = \frac{F}{A}$$

where p = pressure, measured in pascals (Pa),
F = force exerted, measured in newtons (N),
A = area acted on, measured in metres squared (m²).

Therefore, 1 pascal = 1 newton per square metre
$$(1\,Pa = 1\,N\,m^{-2})$$

Worked examples

Example 1

A box rests on a table. The area of the box in contact with the table is 0.25 m². The box exerts a force of 20 N on the table. Calculate the pressure the box exerts on the table.

Solution

$$p = \frac{F}{A} = \frac{20}{0.25} = 80\,Pa$$

Example 2

The pressure exerted by the air in a room is 1.0×10^5 Pa. One of the rectangular walls of the room measures 4.0 m × 3.0 m. Calculate the force exerted on the wall by the air.

Solution

Note: area of a rectangle = (4 × 3) m²

$$p = \frac{F}{A}$$

$$1 \times 10^5 = \frac{F}{(4 \times 3)}$$

$$F = 1 \times 10^5 \times 12 = 1.2 \times 10^6\,N$$

Example 3

An ornamental candle rests on a table. The candle exerts a force of 2.5 N on the table. The pressure of the candle on the table is 318 Pa. Calculate the area of the candle in contact with the table.

Solution

$$p = \frac{F}{A}$$

$$318 = \frac{2.5}{A}$$

$$A = \frac{2.5}{318} = 0.0079\,m^2$$

Physics beyond the classroom

The equation

$$\text{pressure} = \frac{\text{force}}{\text{area}}$$

explains why:

- A sharp knife is easier to cut with than a blunt knife. For the same size of force, a sharp knife exerts a greater pressure than a blunt knife because of the smaller area in contact with the material.

- It is much easier to travel across soft snow wearing skis than walking in snow boots. Skis have a larger surface area than snow boots. The larger area of the skis exerts less pressure on the snow than the boots. Less pressure on the snow means the skis do not sink as far into the snow and so it is easier to move across the soft snow.

- A person can lie on a 'bed of nails'. In this case the force exerted by the person is distributed over the surface area of a large number of nails. Although the surface area of one nail is very small the area provided by the large number of nails means that the pressure exerted on the skin is just below the value that would cause injury.

Why would the person have to be laid carefully on the 'bed of nails'? How can the person get off the 'bed of nails' without injury?

Figure 5.2 Stiletto heels exert high pressure on flooring – sometimes the high pressure exerted by the heels damages the floor

Figure 5.3 Large tyres spread the weight of the vehicle over a large area. This allows the vehicle to travel over soft or muddy ground

Pressure due to matter

- Solid – a solid exerts a pressure on a surface because of the force (its weight) that it exerts on the area in contact with the surface.
- Liquid – a liquid exerts a pressure on the base of its container because of the force (its weight) that it exerts on the area in contact with the surface.
- Gas – it is more difficult to explain how a gas exerts a pressure on a surface. However, consider the following argument:

Kinetic theory assumes that the particles of a gas are moving in all directions (random motion). Evidence for this is provided by the motion of smoke particles in a smoke cell (Figure 5.4).

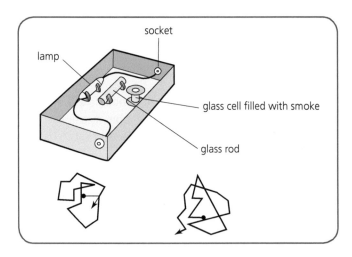

Figure 5.4 A Brownian smoke cell and the random motion of two smoke particles in the cell

The smoke particles have a jerky, irregular motion – this is called Brownian motion. The random motion is caused by the bombardment of the smoke particles by the invisible air particles. The air particles are moving in all directions so that they collide with the smoke particles from all directions and so cause the jerky, irregular motion.

A greatly magnified 'picture' or model of the particles making up a gas in a small container with a moveable piston is shown in Figure 5.5.

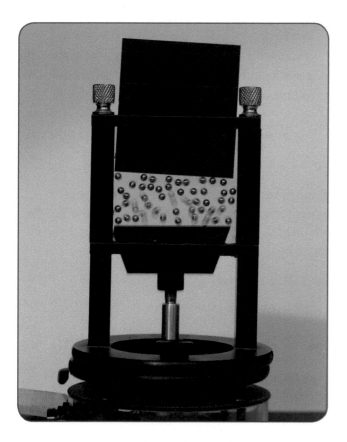

Figure 5.5 'Gas' particles bombarding the walls of a container

In this 'model' each small metal sphere represents a particle of gas. The spheres are made to move rapidly by the vibrating plate at the base. Each time a sphere hits the piston, it exerts a small upwards force on the piston. However, a large number of spheres hit (or bombard) the piston every second. This bombardment is able to give a large enough upwards force to push the piston part of the way up the tube. The pressure on the piston is due to the force exerted by the metal spheres on the area of the piston.

From the above, it can be concluded that the pressure exerted by a gas on the area of a surface it is in contact with is due to the force of the particles that collide with the surface.

Kinetic theory of a gas

The kinetic theory of a gas assumes that the tiny gas particles are in constant random motion. The pressure exerted by a gas on the walls of the container is due to the bombardment of the container walls by the fast-moving gas particles (about $500\,\mathrm{m\,s^{-1}}$ for air at room temperature). The particles hit the walls of the container and rebound, exerting a force on the wall. The pressure is a result of these forces on the wall area.

Pressure depends on the number of collisions on the walls in 1 s and the force of each collision, i.e. pressure depends on particle bombardment.

- A gas is made up of tiny particles.
- The particles of a gas move around in all directions (random motion).
- The particles of a gas collide elastically with each other and with the container walls. Elastically means that the particles do not 'lose' kinetic energy in the collisions and therefore the speed of the particles does not change.
- The pressure exerted by a gas is caused by the particles colliding with the container walls, i.e. due to particle bombardment on the container walls.
- The particle bombardment depends on the force of each collision on the walls and the number of collisions made with the walls every second.

The gas laws

The three things that can change for a fixed mass of gas are:

- pressure
- volume
- temperature.

Pressure and volume

The apparatus shown in Figure 5.6 is used to find the relationship between the pressure exerted by a gas and the volume occupied by the gas, with the gas temperature remaining constant. The gas used in the experiment is air. The mass of the air is fixed (no air is added or leaves the cylinder during the experiment).

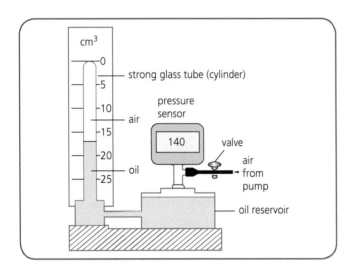

Figure 5.6 Pressure against volume apparatus

The air in the cylinder is first compressed using a foot pump. During the compression the temperature of the air rises as work is being done on it. The air is allowed to return to room temperature before any readings are taken. The air in the cylinder is allowed to expand slowly by opening the valve and the pressure of the air at appropriate volumes noted and recorded, as shown in Table 5.1.

Pressure (kPa)	100	150	200	250	300
Volume (cm³)	24	16	12	9.6	8.0

Table 5.1

The data are plotted on a graph of pressure against volume as shown in Figure 5.7. The resulting graph of pressure exerted by a fixed mass of gas at constant temperature against volume is a smooth curve, which does not cut either axis.

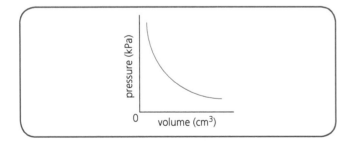

Figure 5.7 Graph of pressure against volume for a fixed mass of gas at constant temperature

However, plotting a graph of pressure against 1/volume as shown in Figure 5.8 gives a straight line through the origin.

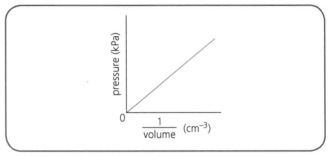

Figure 5.8 Graph of pressure against $\dfrac{1}{\text{volume}}$ for a fixed mass of gas at constant temperature

This means that the pressure exerted by a fixed mass of gas at constant temperature is inversely proportional to the volume of the gas, i.e. $p \propto \dfrac{1}{V}$.

$$\text{pressure} \propto \frac{1}{\text{volume}} \quad or \quad p = \frac{k}{V} \quad \text{where } k \text{ is a constant}$$

$p \times V = k$ This is known as **Boyle's law**.

Let the pressure exerted by a fixed mass of gas at constant temperature be p_1, when its volume is V_1. The volume of the gas is changed to V_2. The pressure now exerted by the gas is p_2. Then:

$$p_1 \times V_1 = \text{constant} = p_2 \times V_2$$

$$p_1 V_1 = p_2 V_2$$

Worked example

Example

The pressure exerted by a fixed mass of gas at constant temperature is 100 kPa. The volume of the gas is 10 m³. The volume of gas is decreased to 5.0 m³. Calculate the new pressure exerted by the gas.

Solution

Note: 100 kPa = 100 × 10³ Pa = 100 000 Pa

$$p_1V_1 = p_2V_2$$

$$100 \times 10^3 \times 10 = p_2 \times 5$$

$$p_2 = \frac{100 \times 10^3 \times 10}{5} = 200 \times 10^3\,\text{Pa}$$

New pressure = 200 kPa

Note that in this law, when the pressure doubles, the volume halves.

Pressure and temperature

The apparatus shown in Figure 5.9 is used to find the relationship between the pressure exerted by a gas and the temperature of a gas while at constant volume. The gas used in this experiment is air. The mass of the gas is fixed (no air enters or leaves the flask during the experiment).

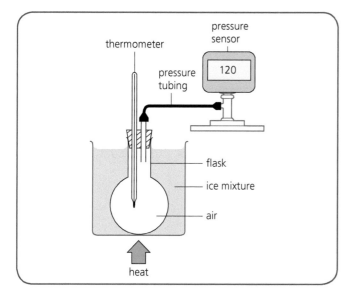

Figure 5.9 Pressure against temperature apparatus

The ice mixture is heated slowly. Readings of pressure and temperature of the gas are taken at regular intervals and recorded, as in Table 5.2.

Pressure (kPa)	102	111	120	129	138
Temperature (°C)	0	25	50	75	100

Table 5.2

The data are plotted on a graph of pressure against temperature as shown in Figure 5.10.

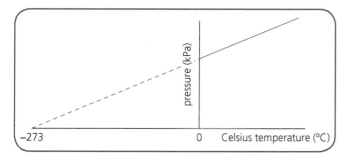

Figure 5.10 Graph of pressure against Celsius temperature at constant volume

The graph of pressure exerted by a fixed mass of gas at constant volume against Celsius temperature is a straight line – but it does not pass through the origin. When drawn backwards (extrapolated) the straight line cuts the temperature axis at −273 °C. At this temperature the pressure exerted by the gas is zero. The temperature of −273 °C is called **absolute zero**.

A new temperature scale called the **Kelvin scale**, which has its zero at absolute zero, is now used. The Kelvin scale differs from the Celsius scale only in the position of its zero point. The Kelvin temperature is obtained by adding 273 to the Celsius temperature.

$$\text{Kelvin temperature} = (\text{Celsius temperature} + 273)$$

$$T\,\text{K} = (t\,°\text{C} + 273)$$

For example:

$$\text{freezing point of water} = 0\,°\text{C} = (0 + 273) = 273\,\text{K}$$

$$\text{boiling point of water} = 100\,°\text{C} = (100 + 273) = 373\,\text{K}$$

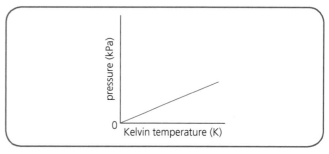

Figure 5.11 Graph of pressure against Kelvin temperature at constant volume

Since the graph of pressure against Kelvin temperature is a straight line through the origin, the pressure exerted by a fixed mass of gas at constant volume is directly proportional to its Kelvin temperature:

$$\text{pressure} \propto \text{Kelvin temperature}$$

$$p = kT \text{ where } k \text{ is a constant}$$

$$\frac{p}{T} = k \qquad \text{This is known as \textbf{Gay-Lussac's law}.}$$

Let the Kelvin temperature of a fixed mass of gas at constant volume be T_1, when the pressure exerted by the gas is p_1. The Kelvin temperature of the gas is changed to T_2. The pressure exerted by the gas is now p_2. Then:

$$\frac{p_1}{T_1} = \text{constant} = \frac{p_2}{T_2}$$

$$\frac{p_1}{T_1} = \frac{p_2}{T_2}$$

Worked example

Example

The pressure exerted by a gas is 1.0×10^5 Pa. The temperature of the gas is 27 °C. The gas is heated at constant volume to a temperature of 127 °C. Calculate the pressure now exerted by the gas.

Solution

Note: temperatures must be in Kelvin

$$\frac{p_1}{T_1} = \frac{p_2}{T_2}$$

$$\frac{1.0 \times 10^5}{(27 + 273)} = \frac{p_2}{(127 + 273)}$$

$$1.0 \times 10^5 \times 400 = p_2 \times 300$$

$$p_2 = \frac{1.0 \times 10^5 \times 400}{300}$$

final pressure $= 1.33 \times 10^5$ Pa

Note that in this law when the pressure of the gas doubles, the Kelvin temperature doubles (but the Celsius temperature does not).

Volume and temperature

The apparatus shown in Figure 5.12 is used to find the relationship between the volume and the temperature of a gas at constant pressure. The gas used in this experiment is air. The mass of the air is fixed (no air is added or leaves the cylinder – the capillary tube – during the experiment). The air in the cylinder is at constant pressure since the air above the bead of mercury is always at atmospheric pressure.

The ice mixture is heated slowly. Readings of volume and temperature of the gas are taken at regular intervals and recorded, as in Table 5.3.

The data are plotted on a graph of volume against temperature (Figure 5.13).

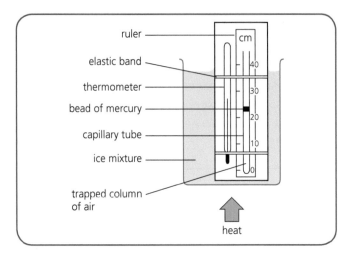

Figure 5.12 Volume against temperature apparatus

Volume of trapped air column (mm³)	58.4	63.8	69.2	74.6	80.0
Temperature (°C)	0	25	50	75	100

Table 5.3

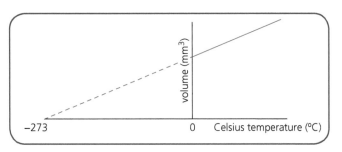

Figure 5.13 Graph of volume against Celsius temperature at constant pressure

The graph of volume of a fixed mass of gas at constant pressure against Celsius temperature is a straight line – but it does not pass through the origin. When extrapolated the straight line cuts the temperature axis at −273°C (absolute zero). At this temperature the volume is zero.

A new temperature scale called the Kelvin scale, which has its zero at absolute zero, is now used (Figure 5.14). The Kelvin scale differs from the Celsius scale only in the position of its zero point.

Since the graph of volume against Kelvin temperature is a straight line through the origin, the volume of a fixed mass of gas at constant pressure is directly proportional to its Kelvin temperature.

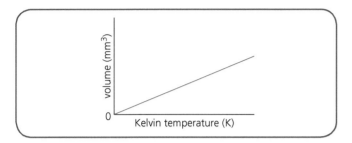

Figure 5.14 Graph of volume against Kelvin temperature at constant pressure

volume ∝ Kelvin temperature

$V = kT$, where k is a constant

$\dfrac{V}{T} = k$ This is known as **Charles's law.**

Let the Kelvin temperature of a fixed mass of gas at constant pressure be T_1, when the volume of the gas is V_1. The Kelvin temperature of the gas is changed to T_2. The volume of the gas is now V_2. Then:

$$\frac{V_1}{T_1} = \text{constant} = \frac{V_2}{T_2}$$

$$\frac{V_1}{T_1} = \frac{V_2}{T_2}$$

Worked example

Example

The volume of a gas is 2.5 m³. The temperature of the gas is 27 °C. The gas is heated at constant pressure until its volume is 5.0 m³. Calculate the temperature of the gas at this volume.

Solution

Note: temperatures must be in Kelvin

$$\frac{V_1}{T_1} = \frac{V_2}{T_2}$$

$$\frac{2.5}{(27 + 273)} = \frac{5.0}{T_2}$$

$$2.5 \times T_2 = 5.0 \times 300$$

$$T_2 = \frac{1500}{2.5} = 600\,\text{K}$$

final temperature = (600 − 273 °C) = 327 °C

Note that, in this law, when the volume of the gas doubles, the Kelvin temperature doubles (but the Celsius temperature does not).

The general gas law

The above laws are experimental laws based on observed behaviour of real gases. However, further experiments show that they are not exact descriptions of real gases. They are, however, good approximations for gases that are well above their boiling points – for example, air at room temperature obeys these laws closely. However, at temperatures of about −150 °C (123 K) the laws do not give a good description of the behaviour of a real gas. Real gases, when cooled sufficiently, change state, i.e. the gas condenses into a liquid.

Mathematically combining Boyle's law and Gay-Lussac's law gives:

$$\frac{pV}{T} = \text{constant} = k \quad \text{or} \quad \frac{p_1 V_1}{T_1} = \frac{p_2 V_2}{T_2}$$

where p = pressure exerted by the gas, measured in pascals (Pa),
V = volume occupied by the gas, measured in cubic metres (m³),
T = temperature of the gas, measured in Kelvin (K).

In some problems different units of pressure and volume are used; the same units are used for both the initial and final values. For example, the initial and final pressures could be measured in atmospheres and the initial and final volumes can be measured in mm³.

(In the pressure against volume and temperature against volume experiments the gas was trapped in a cylinder. The volume of a cylinder = cross-sectional area × height and the cross-sectional area for a cylinder is constant. Therefore, the volume is directly proportional to the height of the cylinder. This means that the volume can also be represented by the height (or length) of the cylinder.)

Only the Kelvin temperature scale can be used when applying the gas laws.

Kelvin temperature = (Celsius temperature + 273)
$$T\,\text{K} = (t\,°\text{C} + 273)$$

Note that:

- 1 degree Celsius does not equal 1 Kelvin.
- A change of 1 degree Celsius is the same as a change of 1 Kelvin.

This means that, for instance, the specific heat capacity of water can be written as $4180\,\text{J}\,\text{kg}^{-1}\,^{\circ}\text{C}^{-1}$ or $4180\,\text{J}\,\text{kg}^{-1}\,\text{K}^{-1}$.

Worked examples

Example 1

The air in a room is at a temperature of 20 °C. A heater is switched on and the temperature of the air rises to 22 °C.
a) Calculate the change in temperature in degrees Celsius.
b) Calculate the change in temperature in Kelvin.

Solution

a) change in temperature, $\Delta T = 22 - 20 = 2\,^{\circ}\text{C}$
b) initial temperature in Kelvin = $(20 + 273) = 293\,\text{K}$

final temperature in Kelvin = $(22 + 273) = 295\,\text{K}$

change in temperature, $\Delta T = 295 - 293 = 2\,\text{K}$

Example 2

A sample of gas occupies a volume of 80 litres. The temperature of the gas is 25°C. The pressure exerted by the gas is 3.0 atmospheres. The temperature of the gas is now doubled while the volume occupied by the gas is halved. Calculate the pressure now exerted by the gas.

Solution

Note: temperatures must be in Kelvin

$$\frac{p_1 V_1}{T_1} = \frac{p_2 V_2}{T_2}$$

$$\frac{3 \times 80}{(25 + 273)} = \frac{p_2 \times 40}{(50 + 273)}$$

$$\frac{240}{298} = \frac{p_2 \times 40}{323}$$

$$240 \times 323 = p_2 \times 40 \times 298$$

$$p_2 = \frac{240 \times 323}{40 \times 298} = 6.5 \text{ atmospheres}$$

Notes on gas pressure

1 Absolute zero ($-273\,^{\circ}\text{C}$ or $0\,\text{K}$) is the temperature at which the pressure of a fixed mass of an ideal gas is zero. At absolute zero the particles are not moving (zero speed) and therefore have no kinetic energy.

2 In the formula

$$\frac{p_1 V_1}{T_1} = \frac{p_2 V_2}{T_2}$$

p_1 and p_2 are the total pressures being exerted on the gas, i.e. including atmospheric pressure.

3 Using the kinetic theory model of matter and Newton's laws (these are discussed in detail in Chapters 9, 10 and 11), gas pressure can be explained in terms of particle bombardment:
   a) particles hit walls of container
   b) particles change direction
   c) velocity of particles changes
   d) particles accelerate
   e) wall of container exerts an unbalanced force on particles (from Newton's second law)
   f) particles exert an unbalanced force on wall of container (Newton's third law)
   g) pressure on wall due to this force (pressure = force/area).

Gas pressure and kinetic theory

- The pressure exerted by a gas on its container is a result of the bombardment of the walls by the fast-moving gas particles. The particles hit the walls and rebound, exerting a force on the wall. The pressure is a result of these forces. Because the number of particles is very large and they are moving very fast, the number of collisions with the walls per second is extremely large. As a result, the series of tiny forces is smoothed out into a steady push on the container walls.

- The particles of the gas collide elastically with each other and the container walls. (Elastically means that during the collisions the kinetic energy of the particles does not change. This means that the speed of the particles does not change.)

- The temperature of a gas depends on the average kinetic energy of the particles. When a gas is heated, the temperature of the gas increases. This means that the average kinetic energy of the gas particles increases and so their speed increases. The gas particles will hit the container walls with more force and more often, so increasing the particle bombardment. This increases the pressure exerted by the gas.

Examples

Consider a fixed mass of gas trapped in a rigid container, which is sealed at one end by a gas-tight movable piston (Figure 5.15). When the piston is stationary the pressure on both sides is the same and is equal to atmospheric pressure.

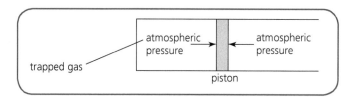

Figure 5.15 A fixed mass of gas trapped in a container by a moveable piston

Pressure and temperature (at constant volume)

The volume of gas is kept constant by preventing the piston from moving.

When the gas is heated, the temperature of the gas will rise. The average kinetic energy of the gas particles will increase and so they will speed up. The particles will hit the container walls harder (greater force) and more often (because they are travelling faster in the same distance) thus increasing the particle bombardment. This causes the pressure exerted by the gas to increase.

As the temperature of the gas increases so does the pressure exerted by the gas as long as the volume is unchanged.

Pressure and volume (at constant temperature)

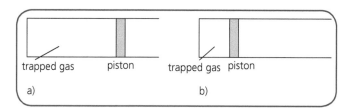

Figure 5.16 a) Before piston is pushed in **b)** After piston is pushed in

When the piston is moved in, the volume occupied by the gas decreases (Figure 5.16). When the gas returns to its original temperature the average kinetic energy of the particles is the same as before, i.e. the particles travel at the same speed. The particles hit the piston with the same force. However, since the volume has decreased, there will be more collisions with the container walls per second (or collisions occur more often). The bombardment on the piston increases and therefore the pressure exerted by the gas increases.

As the volume decreases the pressure exerted by the gas increases provided the temperature is unchanged.

Volume and temperature (at constant pressure)

When the gas is heated, the temperature of the gas will rise. The average kinetic energy of the gas particles will increase so they will speed up. The particles will hit the container walls harder (greater force) and more often (because they are travelling faster in the same distance), thus increasing the particle bombardment. This causes the pressure exerted by the gas on the piston to increase. The piston moves out and the volume occupied by the gas increases. The gas particles have more distance to travel between collisions. There will now be fewer collisions with the container walls per second (or collisions occur less often) and so the particle bombardment will decrease. This cancels out the increase in particle bombardment due to the increase in temperature. The pressure exerted by the gas is unchanged.

Increasing the temperature of the gas increases the volume of the gas as long as the pressure exerted by the gas is unchanged.

Physics beyond the classroom

At normal atmospheric pressure, water boils at 100 °C. However, at pressures higher than atmospheric pressure, water boils at a higher temperature. The steam produced will be at a temperature greater than 100 °C. In a pressure cooker, due to the higher pressure, the temperature of the steam is about 115 °C. The steam is forced through the food, which cooks more rapidly as a result of the higher temperature. This saves energy.

Key facts and physics equations: gas laws and the kinetic model

- Pressure = $\dfrac{\text{force exerted}}{\text{area acted on}}$, i.e. $p = \dfrac{F}{A}$.

- Pressure is measured in pascals (Pa), force exerted is measured in newtons (N) and area acted on is measured in metres squared (m^2).

- The kinetic model explains how the particles of a gas are able to exert pressure due to particle bombardment on the walls of a container.

- Temperature is a measure of the average kinetic energy of the particles making up the substance.

- The general gas law is: $\dfrac{p_1 V_1}{T_1} = \dfrac{p_2 V_2}{T_2}$.

- Pressure is measured in pascals (Pa) or any appropriate unit (e.g. atmospheres) provided the units are the same on both sides of the equation.

- Volume is measured in metres cubed (m^3) or any appropriate unit (e.g. mm^3) provided the units are the same on both sides of the equation.

- Temperature is measured only in Kelvin (K) where $T\,\text{K} = (t\,^\circ\text{C} + 273)$.

- Absolute zero = 0 K (or −273 °C). This is the temperature at which in an ideal gas the particles would have zero kinetic energy, i.e. the particles would be stationary.

End-of-chapter questions

1 A box measures 1.50 m × 0.20 m × 0.30 m. The box exerts a force of 100 N on the ground.
   a) When placed flat on the ground, state the dimensions of the cube that would allow it to:
      i) exert the least pressure on the ground
      ii) exert the greatest pressure on the ground.
   b) Calculate the pressure exerted by the cube in a) i) and a) ii).

2 A rectangular wall of a room measures 3.0 m × 5.0 m. The pressure exerted by the air in a room is 1.0 × 10⁵ Pa. Calculate the force the air particles exert on the wall.

3 A box exerts a force of 50 N on a bench. The pressure exerted by the box on the bench is 250 Pa.
   a) State what is meant by the term *pressure*.
   b) Calculate the area of the box in contact with the bench.

4 a) Change the following temperatures to Kelvin:
      i) 0 °C
      ii) 27 °C
      iii) 100 °C
      iv) 127 °C
      v) −173 °C.
   b) Change the following temperatures to degrees Celsius:
      i) 0 K
      ii) 200 K
      iii) 293 K
      iv) 310 K
      v) 353 K.

5 A compressor pump has a cylinder of volume 0.20 m³. It draws in air at a pressure of 1.0 × 10⁵ Pa. The air is now compressed to a volume of 0.05 m³. Calculate the pressure exerted by the compressed air after the air has returned to its original temperature.

6 A bicycle pump contains a column of air 400 mm long. The pressure exerted by the air is 1.0 atmosphere. The pump handle is pushed in until the pressure exerted by the trapped air rises to 8.0 atmospheres.
   a) Calculate the distance the handle has been pushed in.
   b) What have you assumed about the air?

7 The initial temperature of a sample of gas is 17 °C. The gas is allowed to expand from 1600 m³ to 2000 m³. During the expansion the pressure exerted by the gas remains constant. Calculate the temperature of the gas after the expansion.

8 A balloon contains 5 litres of air at 20 °C. The balloon is placed in a freezer at a temperature of −18 °C. Calculate the volume of the balloon when it is in the freezer.

9 A fixed mass of gas is kept at a constant volume. The temperature of the gas is 27 °C. The pressure exerted by the gas is 75 kPa. The temperature of the gas rises by 10 °C. Calculate the pressure exerted by the gas at this new temperature.

10 A sample of gas in a sealed, rigid container exerts a pressure of 2.5 × 10⁵ Pa. The container is heated and the temperature of the gas rises to 27 °C. The pressure exerted by the gas rises to 3.0 × 10⁵ Pa. Calculate the initial temperature of the gas.

11 A sample of gas is kept in a sealed container. The pressure exerted by the gas is 1.0 × 10⁵ Pa. The sample is heated from 300 K to 400 K. The volume of the container increases from 100 mm³ to 200 mm³. Calculate the pressure now exerted by the gas.

12 This question involves the kinetic theory of a gas.
   a) Explain how the pressure exerted by a fixed volume of gas falls as the temperature of the gas decreases.
   b) State what happens to the particles of a gas when absolute zero is reached.

Exam practice for Chapters 1–5

Information, if required, for use in the following questions can be found on the *Data Sheet* on page 170.

1 a) A circuit is set up as shown in Figure E1.1.

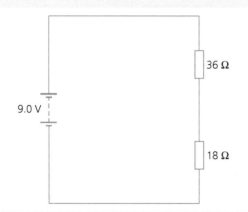

Figure E1.1

   i) Calculate the current drawn from the supply.

   ii) Find the voltage across the 18 Ω resistor.

b) An electrical appliance has three resistors connected as shown in Figure E1.2.

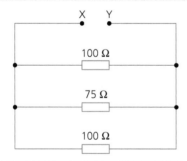

Figure E1.2

   i) Calculate the resistance between points X and Y.

   ii) Points X and Y are connected to a 230V mains supply. Calculate the power rating of the appliance.

2 A set of Christmas tree lights consists of 19 lamps and a resistor, all connected in series. The lamps are labelled 12 V, 3.0 W. The set is connected to the 230 V mains supply and switched on.

a) Why must the 19 lamps and resistor be connected in series?

b) Calculate the current in each of the lamps.

c) Show that the resistance of the resistor is 8.0 Ω.

d) How much charge flows through **each** lamp in 1 minute?

e) Calculate the resistance of each lamp.

3 The power rating of the element of an electric kettle is 2116 W. The element is connected to the 230V mains and switched on for 180s.

a) How much electrical energy is transferred by the element into heat in 180s?

b) When the kettle is switched on, calculate the resistance of the element.

4 A circuit is set up as shown in Figure E1.3.

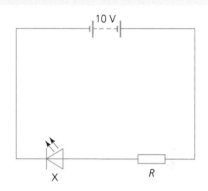

Figure E1.3

a) Name component X.

b) Explain why resistor R is necessary in the circuit.

c) The voltage across component X is 1.8 V. The current in the circuit is 11 mA. Calculate the resistance of resistor R.

5 An electric cooker has two heating plates. Each heating plate is made up of two heating elements, each of resistance 100 Ω. The heating elements are connected in the circuits X and Y as shown in Figure E1.4.

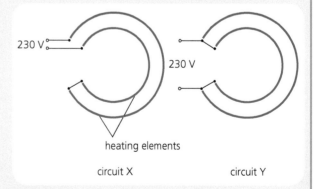

Figure E1.4

a) Which circuit, X or Y, shows the heating elements connected in series?

b) Calculate the resistance of

   i) circuit X

   ii) circuit Y.

c) Both circuits are now switched on. Calculate the current in
   i) circuit X
   ii) circuit Y.

d) Which circuit, X or Y, would heat a pot of soup in the shortest time? Explain your answer.

6 In the circuit shown in Figure E1.5, the variable resistor is adjusted so that the LED is just off when the temperature of the thermistor is 20 °C.

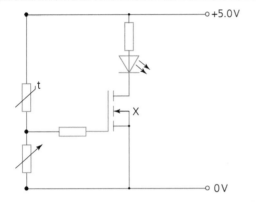

Figure E1.5

a) Name component X.

b) The resistance of the thermistor decreases as its temperature increases. Explain why the LED lights when the temperature of the thermistor rises to 22 °C.

c) The circuit could be altered to warn a gardener of low temperature conditions. What alteration should be made to the circuit diagram to allow the LED to light during low temperatures?

7 A student designs an electronic circuit to switch a lamp on in a room when it gets dark outside. The student's circuit is shown below.

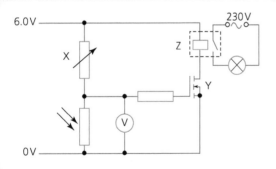

Figure E1.6

a) Name components Y and Z.

b) What happens to the resistance of the LDR when the light level decreases?

c) When the voltmeter reading reaches 2.6 V component Y switches on. Explain how the circuit operates when the light level decreases.

d) Why is a variable resistor chosen for component X rather than a resistor with a fixed value?

8 An ice tray contains 0.15 kg of water. The initial temperature of the water is 20 °C. The ice tray and the water are placed in a freezer. The freezer is at a constant temperature of −19 °C. Show that the freezer has to remove 68.6 kJ of energy from the 0.15 kg of water to change it from water at 20 °C into ice at −19 °C.

9 A heating element is completely immersed in 1.2 kg of water. The element is connected to a 12 V supply and switched on for 23 minutes. In this time the temperature of the water rises by 12.5 °C.

a) Calculate the amount of energy gained by the water.

b) i) Calculate the power rating of the heater.
   ii) State any assumption that you have made in b) i).

c) Calculate the current in the element of the heater.

10 A sample of a solid is at a temperature of 20 °C. The mass of the sample is 0.4 kg. A heater, rated at 100 W, is used to heat the sample for 1100 s. The graph in Figure E1.6 shows how the temperature of the sample varies with time.

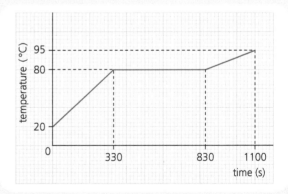

Figure E1.7

a) What is the melting point of the sample?

b) Calculate the specific heat capacity of the sample in the solid state.

c) Calculate the specific latent heat of fusion of the sample.

11 An astronaut, returning from a 'space-walk', re-enters a space station via a door into a vacuum chamber (labelled X).

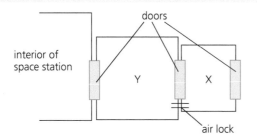

Figure E1.8

Chamber X has a volume of 3.5 m³ and is connected to an inner chamber Y via an airlock. Chamber Y has a volume of 7.0 m³ and is full of air. When the outer door is closed the air lock is opened. Air then rushes into chamber X from Y until the pressure in both chambers is 100 kPa.

a) Calculate the pressure of the air in chamber Y before the airlock was opened.

b) State any assumption that you have made in a).

12 An advertising balloon is filled with 500 m³ of helium gas. The temperature of the helium is 20 °C. During the night the temperature of the helium falls to 5.0 °C. The pressure in the balloon remains unchanged. Calculate the volume occupied by the helium gas at 5.0 °C.

13 A sample of a gas is contained in a sealed rigid container. The pressure exerted by the gas is 1.5×10^5 Pa. The temperature of the gas is 10 °C. The container is heated and the temperature of the gas rises. When the pressure exerted by the gas exceeds 2.2×10^5 Pa the container will break. Calculate the temperature of the gas that will just cause the container to break.

14 A sample of gas is contained in a cylinder that has a smooth-fitting moveable piston. The volume of the sample is 0.0025 m³. The temperature of the sample is −70 °C. The cylinder is heated and the sample of gas is allowed to expand at constant pressure. The gas reaches a temperature of 27 °C. Calculate the volume of the gas at this temperature.

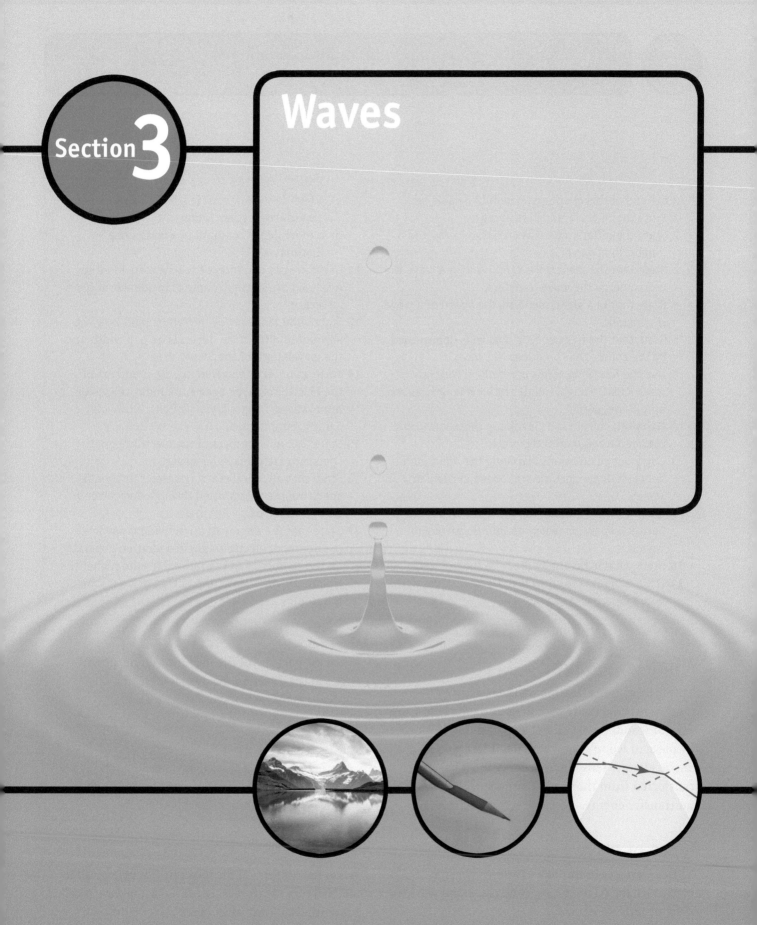

Section 3

Waves

6 Waves and wave phenomena

At the end of this chapter you should be able to:

1 State that a wave transfers energy.
2 State the difference between a transverse and a longitudinal wave.
3 State that the greater the amplitude of a wave the more energy the wave transfers.
4 State that in a given medium, the speed of a wave is constant.
5 State that the frequency of a wave is determined by the source that produces the wave.
6 Use the following terms correctly in context: wave, crest, trough, frequency, wavelength, speed, amplitude, period.
7 Carry out calculations involving the relationship between frequency and period.
8 Carry out calculations involving the relationship between distance, time and speed in problems on waves.
9 Carry out calculations involving the relationship between speed, wavelength and frequency for waves.
10 State what is meant by the refraction of light.
11 Draw diagrams to show the change in direction when light passes from:

a) a less dense material (e.g. air) to a more dense material (e.g. glass, water)
b) a more dense material to a less dense material.
12 Use correctly in context the terms: incident ray, refracted ray, normal, angle of incidence, angle of refraction.
13 State what is meant by diffraction and how the amount of diffraction depends on gap width and the wavelength of the wave.
14 State, in order of wavelength, the members of the electromagnetic spectrum: radio, television, microwaves, infrared, visible light, ultraviolet, X-rays, gamma rays.
15 State that all members of the electromagnetic spectrum are transverse waves.
16 State that all members of the electromagnetic spectrum are transmitted through a vacuum or air at a speed of 3×10^8 m s^{-1}.
17 State that the energy of an individual particle (a photon) of electromagnetic radiation depends on the frequency of the radiation – the higher the frequency, the greater the energy of the particle (photon).

What is a wave?

A piece of wood is floating on the surface of a pond. When waves are produced and pass below the piece of wood, the wood bobs up and down. Since the piece of wood is now moving it has 'gained' kinetic energy. This kinetic energy must have been transferred from the wave. This means that waves can transfer energy.

Types of wave

When a wave passes through a material (the material is often called the medium), the wave causes the particles of the material to vibrate.

When the particles of the medium vibrate at right angles to the direction the wave is travelling in, the wave is called a **transverse wave** (Figure 6.1).

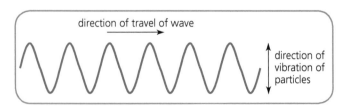

direction of travel of wave

direction of vibration of particles

Figure 6.1 A transverse wave

When the particles of the medium vibrate parallel to the direction the wave is travelling in, the wave is called a **longitudinal wave** (Figure 6.2).

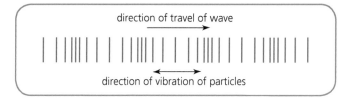

Figure 6.2 A longitudinal wave

Sound (and ultrasound) waves are longitudinal waves. All the other types of wave that you meet in this chapter are transverse waves.

Wave terms

The profile of a transverse wave is shown in Figure 6.3.

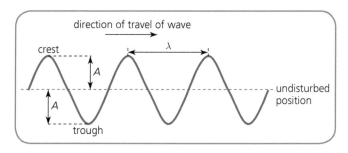

Figure 6.3 Wave terms

The following terms are used to describe a wave.

- **Crest** – the highest point or top of the wave.
- **Trough** – the lowest point or bottom of the wave.
- **Amplitude** (A) – half the height of the wave (from the top or bottom of a wave to the undisturbed position), measured in metres (m). The amplitude of a wave is a measure of how much energy a wave has – the bigger the amplitude, the greater the energy.
- **Wavelength** (λ, lambda) – the distance between two successive similar points (for example from one crest to the next crest), measured in metres (m).
- **Frequency** (f) – the number of waves produced in 1 second, measured in hertz (Hz).

$$\text{frequency} = \frac{\text{number of waves produced}}{\text{time taken to produce the waves}}$$

$$f = \frac{N}{t}$$

- The frequency of a wave is determined by the frequency produced by the source of the waves, i.e. if a source produces waves with a frequency of 20 Hz then this will always be the frequency of these waves. This means that 20 waves are produced in 1 second and 20 waves pass any point in 1 second.

- **Period** (T) – the time taken to produce one wave, measured in seconds (s).

$$\text{period of a wave} = \frac{1}{\text{frequency of the wave}}$$

$$T = \frac{1}{f} \left(\text{or } f = \frac{1}{T} \right)$$

- **Speed of a wave** (v) – the distance travelled by the wave in 1 second. Provided the material remains the same, the speed of a wave is constant.

$$\text{average speed of wave} = \text{speed of wave} = \frac{\text{distance}}{\text{time}}$$

$$v = \frac{d}{t}$$

where v = speed of wave, measured in metres per second (m s^{-1}),
d = distance, measured in metres (m),
t = time, measured in seconds (s).

Wave speed

Consider the transverse wave shown in Figure 6.4.

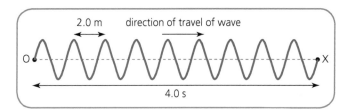

Figure 6.4

The wave takes 4.0 s to travel from O to X.

Between O and X there are 9 complete waves. Each wave has a wavelength of 2.0 m ($\lambda = 2$ m).

Therefore, distance between O and X = 9 × 2 = 18 m.

$$\text{frequency of the waves} = \frac{\text{number of waves}}{\text{time}} = \frac{9}{4} \text{ Hz}$$

Since the medium does not change, then:

$$\text{average speed of wave} = \text{speed of wave} = \frac{\text{distance}}{\text{time}}$$

$$\text{speed of wave} = \frac{9 \times 2}{4}$$

but frequency $= \frac{9}{4}$ and wavelength = 2 m

$$\text{speed of wave} = \text{frequency} \times \text{wavelength}$$
$$v = f \lambda$$

65

The speed of a wave in a given material (medium) can be found using:

$$\text{speed} = \frac{\text{distance}}{\text{time}} \text{ or speed} = \text{frequency} \times \text{wavelength}$$

$$v = \frac{d}{t} \text{ or } v = f\lambda$$

where v = speed of wave, measured in metres per second (m s^{-1}),
d = distance, measured in metres (m),
t = time, measured in seconds (s),
f = frequency of wave, measured in hertz (Hz),
λ = wavelength of wave, measured in metres (m).

Worked examples

Example 1

The speed of water waves in a tank is 0.6 m s^{-1}. The wave generator produces one wave every 0.125 s.
a) What is the frequency of the waves?
b) Calculate the wavelength of the waves.

Solution

a) $f = \frac{1}{T}$

$f = \frac{1}{0.125}$

$f = 8.0\,\text{Hz}$

b) $v = f\lambda$

$0.6 = 8 \times \lambda$

$\lambda = \frac{0.6}{8} = 0.075\,\text{m}$

Example 2

A loudspeaker produces 51 000 sound waves every minute. The wavelength of the waves is 400 mm. Calculate the speed of the waves.

Solution

$\text{frequency} = \dfrac{\text{number of waves produced}}{\text{time taken to produce the waves}} = \dfrac{N}{t}$

$\text{frequency} = \dfrac{51\,000}{60}$ Note: 1 minute = 60 s

$\text{frequency} = 850\,\text{Hz}$

$v = f\lambda$

$v = 850 \times 400 \times 10^{-3}$ Note: 400 mm = 400 × 10^{-3} m = 0.4 m

$v = 340\,\text{m s}^{-1}$

Example 3

A device that emits sound waves is used to find the depth of water below a boat, as shown in Figure 6.5.

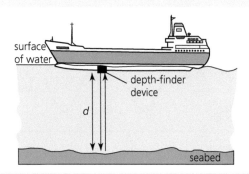

Figure 6.5

The device emits a sound and 0.84 s later an echo is detected. The sound waves travel through the water at a speed of 1500 m s^{-1}. Calculate the depth of the water below the boat.

Solution

Note: 2d is the distance travelled by the sound, i.e. from boat to seabed and back

$v = \dfrac{2d}{t}$

$1500 = \dfrac{2d}{0.84}$

$2d = 1500 \times 0.84 = 1260\,\text{m}$

distance to seabed, $d = \dfrac{1260}{2} = 630\,\text{m}$

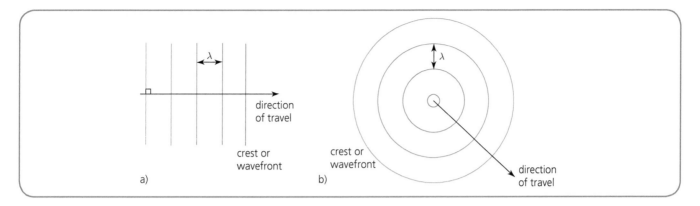

Figure 6.6 a) Plane waves; **b)** circular waves

Representing waves

We sometimes represent waves using lines as shown in Figure 6.6.

These lines represent the crests of the wave. They are one wavelength (λ) apart. The crests are sometimes called **wavefronts**.

Reflection of light

Reflection is illustrated in Figure 6.7.

Figure 6.7 Mountain reflected on the water of a calm lake

Light is a transverse wave. A ray is a line drawn at right angles to the wavefronts of the light. The ray shows the direction of travel of the wave. Figure 6.8 shows a ray of light travelling towards a mirror.

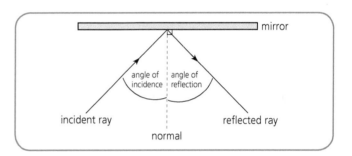

Figure 6.8 Reflection of light at a mirror

- The **incident ray** is the ray that travels towards the mirror.
- The **reflected ray** is the ray that is reflected from the mirror.
- The **normal** is a line at right angles to the mirror.
- The **angle of incidence** is the angle measured between the incident ray and the normal.
- The **angle of reflection** is the angle measured between the reflected ray and the normal.

For reflection:

angle of incidence = angle of reflection

Refraction of light

The effect of refraction is illustrated in Figure 6.9.

Light rays travel through a material, called the medium, in a straight line. When light meets a new medium, some of the light is reflected back and some enters the new medium. When light enters a new medium the speed of the light changes – this is called **refraction** (see Figure 6.10).

Figure 6.9 Refraction makes this pencil look bent

The speed of any wave is given by $v = f\lambda$. When refraction occurs, the speed of the wave changes. However, the frequency of the wave is unchanged (as the frequency of a wave is determined by the frequency produced by the source). This means that the wavelength of the wave must change. When the speed of the wave decreases (frequency does not change) then the wavelength must decrease.

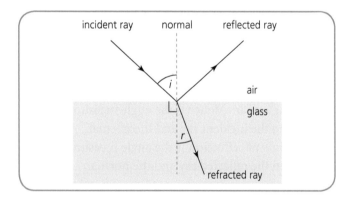

Figure 6.10 Refraction of light at an air–glass boundary

- The **incident ray** is the ray that travels towards the new medium.
- The **refracted ray** is the ray that enters the new medium.
- The **normal** is a line at right angles to the boundary between the two media.
- The **angle of incidence** (i) is the angle measured between the incident ray and the normal.
- The **angle of refraction** (r) is the angle measured between the refracted ray and the normal.

For clarity, in the following diagrams any reflected rays have been omitted.

Figure 6.11 shows a light ray travelling through two glass blocks.

When the light enters the first glass block, the light slows down. When the light leaves the glass block, the light speeds up again. Although the light ray has not changed direction (as it is already travelling along the normal), the light has been refracted (twice) since the speed of the light changed (twice).

On entering the second glass block, the light slows down (refracts) and changes direction by bending towards the normal. On leaving the glass the light speeds up (refracts) and changes direction by bending away from the normal. For a parallel-sided block, the rays in the air are parallel.

Figure 6.12 shows three light rays being refracted by a semicircular glass block.

As the angle of incidence increases, the angle of refraction also increases but in each case the angle of refraction is smaller than the angle of incidence. In each case, as the refracted light slows down the refracted ray is bent towards the normal.

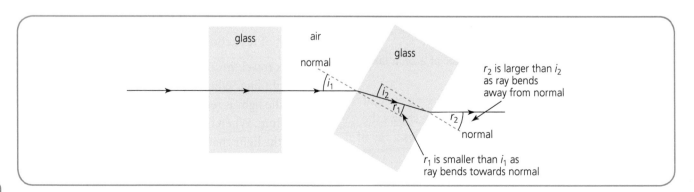

Figure 6.11 Refraction of light at rectangular glass blocks

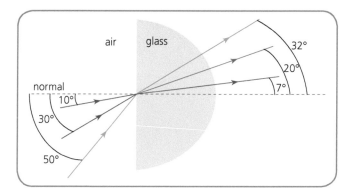

Figure 6.12 Refraction of light at a semicircular glass block

Figure 6.13 shows the effect of a triangular prism on a ray of red light. When the light enters the glass, it slows down and the ray bends towards the normal. When the light enters the air again, it speeds up and the ray bends away from the normal.

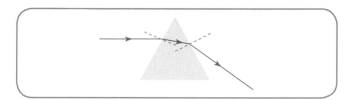

Figure 6.13 Refraction of light through a triangular prism

Knowing the path taken by a light ray through a triangular prism can be useful as this shape can be used to form convex and concave lenses.

Refraction of light: convex lenses

Figure 6.14 shows a simple converging lens that has been made from two triangular prisms and a rectangular block.

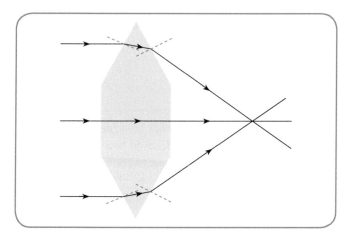

Figure 6.14 A simple convex lens

Smoothing the edges of the shapes gives the more conventional converging (convex) lens shapes shown in Figure 6.15.

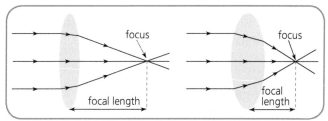

Figure 6.15 'Thin' and 'thick' convex lenses

- A converging (convex) lens brings rays of light to a focus.
- The distance from the centre of the lens to the focus of the lens is called the **focal length**.
- A thick (more curved) lens has a shorter focal length than a thin lens.

Refraction of light: concave lens

Figure 6.16 shows a simple diverging lens that has been made from two triangular prisms and a rectangular block.

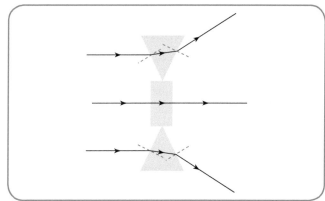

Figure 6.16 A simple diverging lens

Smoothing the edges of the shapes gives the more conventional diverging (concave) lens shape shown in Figure 6.17. A diverging (concave) lens spreads rays of light out.

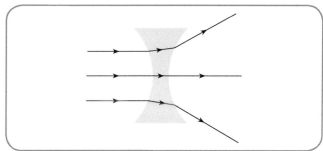

Figure 6.17 A concave lens

Rules for refraction

1 Light travelling from air to glass (Figure 6.18)

- The light slows down.
- The light ray bends towards the normal – the ray makes a smaller angle with the normal.

2 Light travelling from glass to air (Figure 6.19)

- The light speeds up.
- The light ray bends away from the normal – the ray makes a bigger angle with the normal.

In general:

- When light travels from air to any other material it slows down.
- When light travels from any material to air it speeds up.

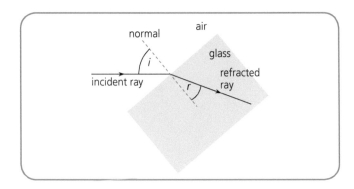

Figure 6.18

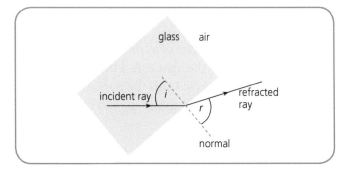

Figure 6.19

Physics beyond the classroom

Refraction can be used to check the quality (density) of liquids using a device called a refractometer (Figure 6.20). This consists of two triangular containers. One part contains a sample of the pure liquid. The other part contains a sample of the test liquid.

If the two liquids have the same density, a ray of light will pass straight through (horizontally).

If the two liquids have different densities the ray of light in the test liquid will be deflected either up or down.

A photo-detector can be used to detect the direction of the ray of light.

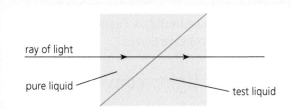

Figure 6.20 A refractometer

Physics beyond the classroom

Opticians use convex and concave lenses to correct some sight defects.

Long sight

A long-sighted person can see distant objects clearly (this means the lens of the eye can be made thin enough). However, objects quite close to the eye appear blurred.

This is because the eye lens cannot be made thick enough and light from a near object is focused beyond the retina. A converging (convex) lens is used to correct this defect. The convex lens increases the bending of the light rays before they enter the eye lens (Figure 6.21). This allows the eye lens to focus the rays of light onto the retina and so the near object is seen clearly.

Short sight

A short-sighted person can see near objects clearly (this means the lens of the eye can be made thick enough). However, distant objects appear blurred. This is because the eye lens cannot be made thin enough and light from a distant object is focused in front of the retina. A diverging (concave) lens is used to correct this defect. The concave lens spreads the rays of light out more before they enter the eye lens (Figure 6.22). This allows the eye lens to focus the rays of light onto the retina and so the distant object is seen clearly.

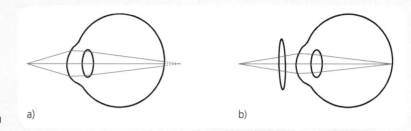

a)   b)

Figure 6.21 a) Near object is blurred; **b)** a convex lens corrects long sight

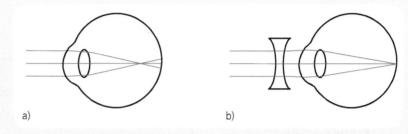

a)   b)

Figure 6.22 a) Distant object is blurred; **b)** a concave lens corrects short sight

Diffraction of waves

When waves pass round an obstacle or pass through a gap, the waves bend. The bending of the waves is called **diffraction**.

Figure 6.23 shows the diffraction of waves of different wavelengths at the same obstacle.

The amount of diffraction (bending) depends on the wavelength of the waves: the longer the wavelength (the lower the frequency, since the speed of the waves does not change and $v = f\lambda$), the greater the amount of diffraction.

Diffraction also occurs when waves pass through a gap (Figure 6.24).

In diagrams X and Y, the waves have the same wavelength. The waves in diagram Y are diffracted more than those in diagram X. The amount of diffraction increases as the width of the gap gets smaller (wavelength being constant).

In diagrams X and Z the width of the gap is the same. The waves in diagram Z are diffracted more than those in diagram X. The amount of diffraction increases as wavelength increases (gap being constant).

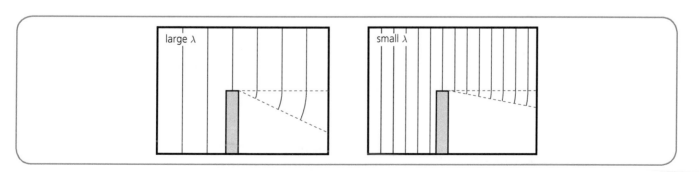

Figure 6.23 Diffraction at an obstacle by long and short wavelengths

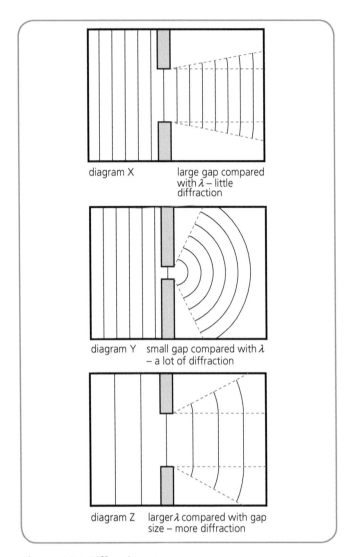

diagram X   large gap compared with λ – little diffraction

diagram Y   small gap compared with λ – a lot of diffraction

diagram Z   larger λ compared with gap size – more diffraction

Figure 6.24 Diffraction at a gap

The amount of diffraction of a wave depends on the size of the gap and the wavelength of the waves:

- For a given wavelength the amount of diffraction increases as the gap size decreases.
- For a given gap size the amount of diffraction increases as the wavelength increases.

Diffraction of radio and television waves

Diffraction applies to all waves. Long wavelength (low frequency) radio waves are diffracted more than short wavelength radio waves (high frequencies).

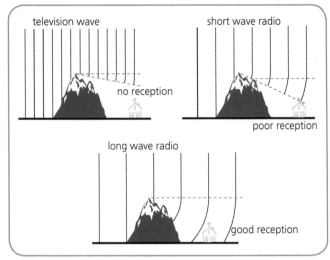

Figure 6.25 Diffraction of television and radio waves

Television waves have a shorter wavelength than radio waves and are diffracted less. This means that television waves are more difficult to pick up in certain areas such as a glen or valley (Figure 6.25).

With the long wavelength (LW) radio signal, the waves diffract more. More of the signal reaches the aerial so a stronger signal is received and there is better reception.

The electromagnetic spectrum

Energy is transferred through space by a family of waves called the **electromagnetic spectrum** (Figure 6.26).

The electromagnetic spectrum consists of the following waves:

Longest wavelength	Radio waves	Lowest frequency
	Television waves	
	Microwaves	
	Infrared radiation	
	Visible light	
	Ultraviolet radiation	
Shortest wavelength	X-rays	Highest frequency
	Gamma rays	

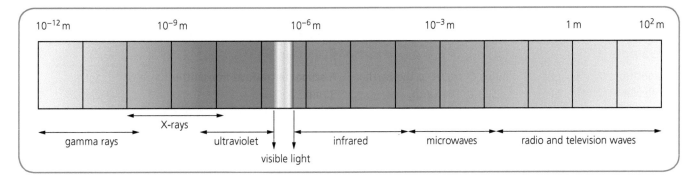

| 10^{-12} m | 10^{-9} m | 10^{-6} m | 10^{-3} m | 1 m | 10^2 m |

gamma rays X-rays ultraviolet visible light infrared microwaves radio and television waves

Figure 6.26 The electromagnetic spectrum

The waves have the following properties:

- They are all transverse waves.
- They all transfer energy.
- They all travel at a speed of 3×10^8 m s^{-1} (300 000 000 m s^{-1}) in a vacuum (and in air).

How the different members of the electromagnetic spectrum can be detected and some of their uses (and dangers) are given below.

Radio and television waves

- Radio and television waves are detected by an aerial and a receiver.
- Very high frequency signals (VHF) are used for television and for FM radio.
- Long wavelength radio signals can be diffracted around hills and over the horizon. They have a very long range.
- Very low frequency signals are able to penetrate into the sea and are used for submarine communication.

Microwaves

- Microwaves are detected by an aerial and a receiver.
- Their very short wavelength means that they undergo little diffraction. They are ideal for line-of-sight communication such as radar, for satellite broadcasting and for mobile phones.

Infrared radiation

- Infrared radiation is detected by a photodiode and meter.
- It is emitted by any hot object but is invisible to the human eye.
- It is used for night 'vision' and in the treatment of muscle injuries.

Visible light

- Visible light is detected by the retina of the human eye and a photodiode and meter.
- It is emitted when electrons move between energy levels in an atom.
- It enables us to see things.

Ultraviolet radiation

- Ultraviolet radiation is detected by fluorescent materials and photographic film.
- It is emitted during electrical discharges such as sparks or lightning. It is also emitted by stars such as our Sun.
- It is used to treat some skin conditions (but over exposure can cause skin cancer), to detect forged bank notes and for 'labelling' equipment with invisible security markings.

X-rays

- X-rays are detected by photographic film and an X-ray intensifier.
- They are emitted when high-speed electrons are slowed down very quickly.
- They are used to 'see' inside the human body (e.g. an X-ray of a broken bone).

Gamma rays

- Gamma rays are detected by a Geiger–Müller tube and counter and photographic film.
- They are emitted by radioactive nuclei.
- They are used as a radioactive tracer and in radiotherapy (see Chapter 7).

Worked examples

Example 1

Light from the Sun takes 8 minutes to travel to the Earth. Calculate the distance from the Sun to the Earth.

Solution

Note: electromagnetic waves travel at $3 \times 10^8 \, \text{m s}^{-1}$ in a vacuum and in air

$$v = \frac{d}{t}$$

$$3 \times 10^8 = \frac{d}{(8 \times 60)}$$

$$d = 3 \times 10^8 \times 480$$

$$d = 1.44 \times 10^{11} \, \text{m}$$

Example 2

Radio Scotland sends out waves with a frequency of 810 kHz. Calculate the wavelength of the radio waves transmitted.

Solution

Note: electromagnetic waves travel at $3 \times 10^8 \, \text{m s}^{-1}$ in air

Note: $810 \, \text{kHz} = 810 \times 10^3 \, \text{Hz} = 810\,000 \, \text{Hz}$

$$v = f\lambda$$

$$3 \times 10^8 = 810 \times 10^3 \times \lambda$$

$$\lambda = \frac{3 \times 10^8}{810 \times 10^3} = 370 \, \text{m}$$

Example 3

A school microwave transmitter has a wavelength of 30 mm.
a) Calculate the frequency of the microwaves.
b) How long will it take the microwaves to travel 1.5 m?

Solution

a) Note: electromagnetic waves travel at $3 \times 10^8 \, \text{m s}^{-1}$ in air

Note: $30 \, \text{mm} = 30 \times 10^{-3} \, \text{m} = 0.030 \, \text{m}$

$$v = f\lambda$$

$$3 \times 10^8 = f \times 30 \times 10^{-3}$$

$$f = \frac{3 \times 10^8}{30 \times 10^{-3}} = 1 \times 10^{10} \, \text{Hz}$$

b) Note: electromagnetic waves travel at $3 \times 10^8 \, \text{m s}^{-1}$ in air

$$v = \frac{d}{t}$$

$$3 \times 10^8 = \frac{1.5}{t}$$

$$t = \frac{1.5}{3 \times 10^8} = 5 \times 10^{-9} \, \text{s}$$

Energy and frequency

The members of the electromagnetic spectrum have been described as waves. However, physicists also use a model that considers electromagnetic radiation to be a stream of individual particles of energy called **photons**. The amount of energy carried by each photon depends on the frequency of the radiation emitted by the source. The higher the frequency of the radiation emitted, the greater the energy the photon has.

Violet light has a higher frequency than red light. This means that photons of violet light have more energy than photons of red light.

The idea of photons of radiation is another model that is used to describe electromagnetic radiation.

Electromagnetic waves are transverse waves and cause the particles of the medium to vibrate at right angles to the direction of travel of the wave.

Light can pass through a vacuum (no particles) – how can this be? Maybe a single model cannot explain all phenomena we meet in physics.

Key facts and physics equations: waves and wave phenomena

- All waves transfer energy.
- When the particles of the medium vibrate at right angles to the direction the wave is travelling in, the waves are called transverse waves.
- When the particles of the medium vibrate parallel to the direction the wave is travelling in, the waves are called longitudinal waves.
- The greater the amplitude of a wave the more energy the wave transfers.
- The frequency of a wave is determined by the source that produces the wave.
- Frequency $= \dfrac{1}{\text{period}}$, i.e. $f = \dfrac{1}{T}$
- The speed of a wave is constant in a given medium.
- Speed of a wave = frequency of wave × wavelength of wave, i.e. $v = f\lambda$
- Speed of a wave $= \dfrac{\text{distance}}{\text{time}}$, i.e. $v = \dfrac{d}{t}$
- Speed is measured in metres per second ($\mathrm{m\,s^{-1}}$), frequency in hertz (Hz), wavelength in metres (m), distance in metres and time in seconds.
- Refraction is the change in speed of light as the light passes from one medium into another.

- The normal is a line at right angles to the surface of the material.
- The angle of incidence is the angle between the incident ray and the normal.
- The angle of refraction is the angle between the refracted ray and the normal.
- When light passes from a less dense medium into a more dense medium, e.g. from air into glass, the light slows down and the ray of light bends towards the normal.
- When light passes from a more dense medium into a less dense medium, e.g. from glass into air, the light speeds up and the ray of light bends away from the normal.
- Diffraction is the bending of waves at a gap or obstacle. The amount of diffraction increases as the gap width decreases and the wavelength of the wave increases.
- The members of the electromagnetic spectrum are: radio waves, television waves, microwaves, infrared radiation, visible light, ultraviolet radiation, X-rays, gamma rays.
- All members of the electromagnetic spectrum are transverse waves and are transmitted through a vacuum or air at a speed of $3 \times 10^{8}\,\mathrm{m\,s^{-1}}$.
- The energy of an individual particle (a photon) of electromagnetic radiation depends on the frequency of the radiation – the higher the frequency, the greater the energy of the particle (photon).

End-of-chapter questions

1   Two types of wave are shown in Figure 6.27.

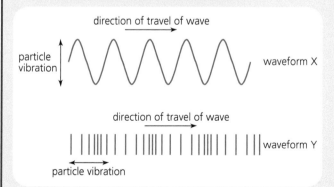

Figure 6.27

   a) Which waveform, X or Y, represents a transverse wave?
   b) Which waveform, X or Y, represents a wave of gamma radiation?

2   Calculate the value of each missing quantity in the table below.

Period	Frequency	Wavelength	Speed
0.2 s	(a)	3.0 m	(b)
(c)	10 Hz	(d)	25 m s⁻¹
0.125 s	(e)	4.0 m	(f)
2.5 s	(g)	(h)	800 mm s⁻¹

3 Figure 6.28 represents a wave.

0.5 m

X

Y

2.0 m

Figure 6.28

The wave travels from X to Y in 0.2 s. For this wave calculate:

a) the amplitude
b) the frequency
c) the wavelength
d) the speed.

4 The depth of a swimming pool is constant. At one end of the pool there is a wave-making machine. The length of the pool is 30 m. Waves take 20 s to travel the length of the pool. The waves have a wavelength of 3.0 m.

a) Calculate the speed of the waves.
b) Calculate the frequency of the waves.
c) The frequency of the wave-making machine is doubled. What effect does this have on the speed and wavelength of the waves in the pool?

5 Refraction of light takes place at the glass block as shown in Figure 6.29.

Figure 6.29

a) What is meant by the refraction of light?
b) Copy the diagram and mark in and label the normal.
c) On your diagram label the angle of incidence as *i* and the angle of refraction as *r*.

6 When light enters a more dense medium the light _____A_____ and the ray _____B_____ the normal – this is known as _____C_____. When light enters a less dense medium the light _____D_____ and the ray _____E_____ from the normal.

Match each letter with the correct word or words from the following list:
- bends away
- bends towards
- reflection
- refraction
- slows down
- speeds up.

7 A ray of light is incident on a rectangular glass block as shown in Figure 6.30.

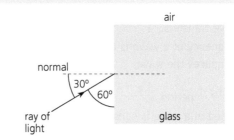

air

normal

30°

60°

ray of light

glass

Figure 6.30

a) State the value of the angle of incidence.
b) Copy the diagram and complete it to show the path taken by the ray inside the glass block.

8 A ray of light travels from glass into water as shown in Figure 6.31.

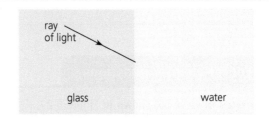

ray of light

glass

water

Figure 6.31

The speed of light in the glass is slower than the speed of light in the water.

a) Copy the diagram and mark in and label the normal at the glass/water boundary.
b) Complete your diagram to show the path taken by the ray in the water.

9 Figure 6.32 shows a ray of light travelling from medium 1 to medium 2.

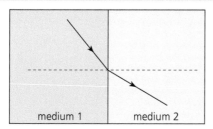

Figure 6.32

  a) When the light enters medium 2 from medium 1, does it slow down or speed up?

  b) Measure
    i) the angle of incidence
    ii) the angle of refraction.

10 Figure 6.33 shows a fisherman on the bank of a river. The fisherman sees a fish in the water.

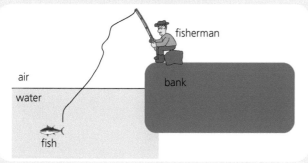

Figure 6.33

Copy the diagram. On your diagram draw a ray of light going from the fish to the fisherman's eye.

11 A man in a house, surrounded by hills, wishes to listen to the radio. The radio can be tuned to two wavebands, LW (150 to 270 kHz) or MW (540 to 1600 kHz). Which of these wavebands is likely to provide the better reception? Explain your answer.

12 a) The electromagnetic spectrum below is in order of increasing frequency.

Radio waves	Television waves	Micro- waves	P	Visible light	Ultra- violet	Q	Gamma rays

    i) Name the missing radiations P and Q.
    ii) Name a detector for ultraviolet radiation.

  b) State **three** properties that members of the electromagnetic spectrum have in common.

13 The wavelength of a source of blue light is 4.80×10^{-7} m.
  a) Calculate the frequency of this light.
  b) The wavelength of a source of red light is 6.44×10^{-7} m. Is the frequency of the red light greater than, equal to or smaller than the blue light in part a)? Explain your answer.

14 A radio signal is broadcast from the Earth to the Moon. The distance between the Earth and the Moon is 384 400 km. Calculate the time taken for the signal to reach the Moon.

Radiation

7 Effects of nuclear radiation

At the end of this chapter you should be able to:

1 Describe a simple model of the atom which includes protons, neutrons and electrons.
2 State what is meant by an alpha particle, beta particle and gamma ray.
3 State that radiation energy may be absorbed in the medium through which it passes.
4 State the approximate range through air, and the absorption by materials, of alpha, beta and gamma radiation.
5 Explain the term 'ionisation'.
6 State that alpha particles are more ionising than beta particles or gamma rays.
7 Describe how one of the effects of radiation is used in a detector of radiation.
8 State that radiation can kill living cells or change the nature of living cells.
9 State that the absorbed dose is the energy absorbed per unit mass of the absorbing material.
10 State that the gray (Gy) is the unit of absorbed dose and that 1 gray is 1 joule per kilogram.
11 Carry out calculations involving the relationship between absorbed dose, energy absorbed and mass of absorbing material.
12 State that a radiation weighting factor is given to each kind of radiation as a measure of its biological effect.

13 State that the equivalent dose is the product of absorbed dose and radiation weighting factor and is measured in sieverts (Sv).
14 Carry out calculations involving the relationship between equivalent dose, absorbed dose and radiation weighting factors.
15 State that equivalent dose rate is the equivalent dose per unit time.
16 Carry out calculations involving the relationship between equivalent dose rate, equivalent dose and time.
17 State that the risk of biological harm from an exposure to radiation depends on:
   a) the absorbed dose
   b) the type of ionising radiation, e.g. α, β, γ, slow neutron
   c) the body organs or tissue exposed.
18 Describe factors affecting the background radiation level.
19 Describe the safety procedures necessary when handling radioactive substances.
20 Identify the radioactive hazard sign and state where it should be displayed.
21 State that exposure to radiation is reduced by shielding, by limiting the time of exposure and by increasing the distance from a source.
22 Describe one medical use of radiation based on the fact that radiation can destroy cells.
23 Describe one use of radiation based on the fact that radiation is easy to detect.

Ionising radiation

A simple model of the atom

All solids, liquids and gases are made up of atoms. An atom consists of a positively charged centre or **nucleus** surrounded by a 'cloud' of rapidly revolving negative charges called **electrons**. The nucleus is made of particles called **protons** (positively charged) and **neutrons** (uncharged). Compared with the size of the atom the nucleus is very small.

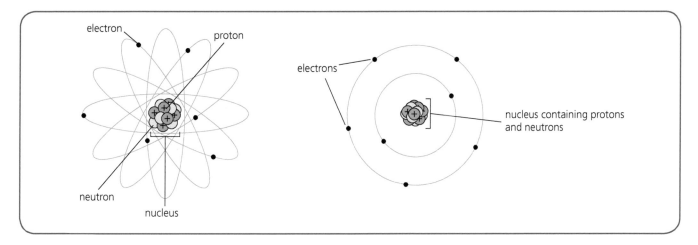

Figure 7.1 Models of the atom

In an uncharged atom the number of protons in the nucleus is equal to the number of electrons orbiting the nucleus.

Types of radiation

The three types of nuclear radiation are:

- Alpha (α) particles – these are helium nuclei. An alpha particle is made up of two protons and two neutrons and is positively charged.
- Beta (β) particles – these are high-speed electrons. A beta particle travels at about 90% of the speed of light (about $2.7 \times 10^8\,\text{m}\,\text{s}^{-1}$) and is negatively charged.
- Gamma (γ) rays – these are members of the electromagnetic spectrum. They travel at $3 \times 10^8\,\text{m}\,\text{s}^{-1}$ in air and are transverse waves. A gamma ray is a burst of energy.

Most materials are made up of atoms in which the nuclei are stable. However, radioactive materials are made up of atoms that have unstable nuclei. These unstable nuclei may become stable when the nucleus disintegrates (or decays) by emitting alpha, beta or gamma radiation. When alpha, beta or gamma radiation is emitted from a nucleus, energy is transferred from the nucleus.

Absorbing alpha, beta and gamma radiation

When alpha particles, beta particles or gamma rays pass through a material, they **transfer energy** to the material

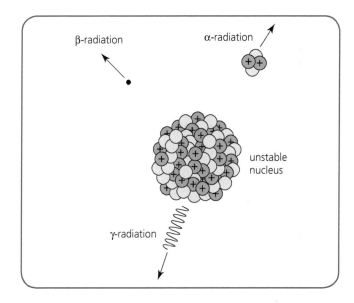

Figure 7.2 Types of radiation that can be emitted from an unstable nucleus

due to interactions or collisions with the atoms that make up the material. Eventually the radiations transfer so much energy due to these collisions that they can penetrate no further through the material – they are said to be absorbed.

Alpha radiation, being relatively large (in atomic dimensions) and consisting of charged particles, undergoes many more interactions or collisions with the atoms of a material in a short distance than either beta or gamma radiation. As a result, alpha particles are absorbed by a few centimetres of air or a thin sheet of paper.

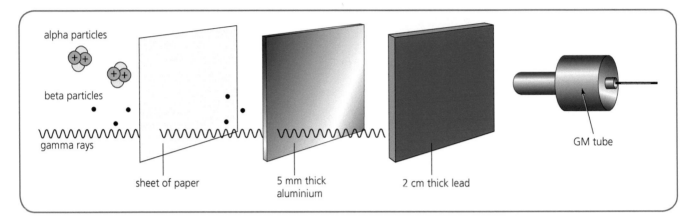

Figure 7.3 Absorption of alpha, beta and gamma radiations

Beta radiation consists of electrons. An electron has a smaller size and charge than an alpha particle, and because of this beta radiation is more penetrating. Beta particles are absorbed by a few metres of air or a few millimetres of aluminium.

Gamma radiation, being a wave, is the most penetrating of the three radiations. Gamma rays require many kilometres of air or a few centimetres of lead (a very dense material) to be absorbed.

The material that the radiation passes through absorbs the energy of the radiation. The amount of energy that is absorbed depends on:

- the type of radiation
- the thickness of the absorbing material
- the type of absorbing material.

Ionisation

When radiation passes through an absorbing material, electrons can be removed from the atoms or molecules in the absorbing material. This process is called **ionisation**. During ionisation, energy is transferred from the radiation to the absorber.

When an electron is removed from an atom, the atom becomes positively charged. The positively charged atom and the 'free' electron are called **ions**. The charged atom is a positive ion and the electron is a negative ion.

Alpha radiation produces the most ionisation. The process of ionisation by an alpha particle is shown in Figure 7.4.

Alpha, beta and gamma radiation are often called **ionising radiations**.

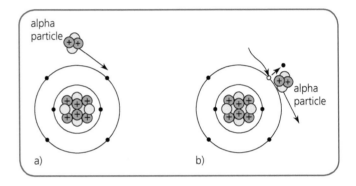

Figure 7.4 Ionisation of a neutral atom by an alpha particle: **a)** alpha particle approaching neutral atom; **b)** alpha particle has passed by having created an ion pair

Ionisation is the breaking up of a neutral atom into positive and negative pieces.

Detecting radiation

Three effects of radiation on non-living things are as follows:

- Radiation fogs (blackens) photographic film.
- Radiation causes ionisation.
- Radiation causes scintillations – causes light to be emitted from certain materials.

Effect of radiation on living things

When living tissue receives a dose of radiation, ionisation takes place within the tissue. This can disturb the way the cells in the tissue operate and can cause the cells to become cancerous.

The film badge

A radiation film badge contains a small piece of photographic film behind thicknesses of different absorbers.

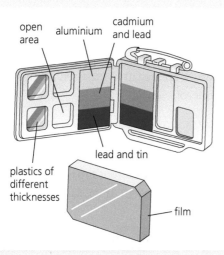

Figure 7.5 A film badge

When the film is developed, the amount of fogging (how black the film is) gives an indication of the amount and type of radiation exposure at the different parts of the badge. The blacker the film, the greater the exposure to radiation.

The Geiger counter

The Geiger–Müller tube (GM tube) and counter makes use of the ionising properties of alpha, beta and gamma radiations.

The GM tube is a hollow cylinder filled with a gas at low pressure. The tube has a thin window, made of a substance called mica, at one end. There is a central electrode inside the tube. A voltage supply is connected across the casing of the tube and the central electrode as shown in Figure 7.6.

When alpha, beta or gamma radiation enters the tube, the radiation produces ions in the gas. The ions created produce a current in the tube for a short time. This current produces a voltage pulse. The voltage pulses are counted and displayed on the counter screen. Each voltage pulse corresponds to one alpha particle, beta particle or gamma ray entering the GM tube.

The scintillation counter: the gamma camera

Some substances such as zinc sulfide fluoresce when they absorb radiation, i.e. they are able to change the energy they absorb into tiny bursts of light. The flashes of light are called scintillations. Scintillations can be observed by the naked eye or they can be counted using a light detector and an electronic counting circuit.

The gamma camera is used to detect gamma radiation. Some patients are given drugs combined with chemicals which emit gamma radiation (see radioactive tracers on page 89). The gamma radiation emitted causes scintillations when they reach the crystal in the gamma camera. These flashes of light are used to build up a picture of the organs inside the human body.

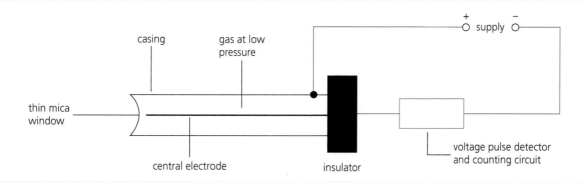

Figure 7.6 A Geiger–Müller tube and counter

Radiation can kill living cells or it can change the nature of the living cell. This can cause problems in the human body.

Effects of radiation on body tissue

Alpha radiation

Alpha radiation produces a lot of ionisation within a short distance in body tissue. Alpha radiation outside the body is absorbed by the skin (it cannot penetrate the skin) and so no damage is done inside the human body. However, a source of alpha radiation inside the body would produce a large amount of ionisation in the affected body tissue, which would be extremely dangerous.

Beta radiation

Beta radiation is absorbed by about 10 mm of body tissue and damage will be caused to that tissue. A source of beta radiation inside the body would cause internal organs to be damaged.

Gamma radiation

Gamma radiation can pass straight through the body. Body tissue can be damaged whether the source of gamma radiation is inside or outside the body.

Other sources

X-rays: X-rays are similar to gamma rays. They are able to pass through the body. Body tissue can be damaged by X-rays.

Neutrons ('fast' and 'slow'): Neutrons are produced during nuclear fission (see Chapter 8). Fast neutrons have more kinetic energy than slow neutrons. Neutrons, being uncharged, can be more penetrating than gamma rays despite being relatively large (in atomic dimensions). Neutrons are able to pass through the body but, during interactions with body tissue, considerable biological damage results.

Absorbed dose

When a radioactive source decays, the radiation emitted has energy. When this energy is absorbed by a material, the material is said to absorb a dose of radiation, called the **absorbed dose**. The absorbed dose can cause damage to the material.

Absorbed dose is the energy absorbed per unit mass (1.0 kg) of the absorbing material.

$$\text{absorbed dose} = \frac{\text{energy absorbed}}{\text{mass of absorbing material}}$$

$$D = \frac{E}{m}$$

where D = absorbed dose, measured in grays (Gy),
E = energy absorbed, measured in joules (J),
m = mass of absorbing material, measured in kilograms (kg).

This means that 1 gray = 1 joule per kilogram ($1\,\text{Gy} = 1\,\text{J}\,\text{kg}^{-1}$).

Absorbed doses are usually measured in milliGrays (mGy) or microGrays (μGy).

($1\,\text{mGy} = 10^{-3}\,\text{Gy} = 0.001\,\text{Gy}$ and $1\,\mu\text{Gy} = 10^{-6} = 0.000001\,\text{Gy}$)

Absorbed doses from different types of radiation *cannot* be added together (as different types of radiation do different amounts of harm to the same material).

Worked examples

Example 1

A sample of tissue absorbs 0.64 mJ of energy. The mass of the tissue is 55 g. Calculate the absorbed dose for this tissue.

Solution

Note: $0.64\,\text{mJ} = 0.64 \times 10^{-3}\,\text{J} = 0.00064\,\text{J}$
and $55\,\text{g} = 55 \times 10^{-3}\,\text{kg} = 0.055\,\text{kg}$

$$D = \frac{E}{m} = \frac{0.64 \times 10^{-3}}{55 \times 10^{-3}} = 0.0116 = 0.012\,\text{Gy}$$

Example 2

A technician's hand is exposed to an absorbed dose of 50 μGy. The hand has a mass of 450 g. Calculate the energy absorbed by the hand.

Solution

Note: $50\,\mu\text{Gy} = 50 \times 10^{-6}\,\text{Gy} = 0.000050\,\text{Gy}$
and $450\,\text{g} = 450 \times 10^{-3}\,\text{kg} = 0.450\,\text{kg}$

$$D = \frac{E}{m}$$

$$50 \times 10^{-6} = \frac{E}{450 \times 10^{-3}}$$

$$E = 50 \times 10^{-6} \times 450 \times 10^{-3}$$

$$E = 2.25 \times 10^{-5}\,\text{J}$$

Example 3

A finger absorbs 0.24 μJ of radiation during an X-ray. The absorbed dose, for the finger, is 4.2 μGy. Calculate the mass of the finger.

Solution

Note: $4.2\,\mu Gy = 4.2 \times 10^{-6}\,Gy = 0.000\,004\,2\,Gy$
and $0.24\,\mu J = 0.24 \times 10^{-6}\,J = 0.000\,000\,24\,J$

$$D = \frac{E}{m}$$

$$4.2 \times 10^{-6} = \frac{0.24 \times 10^{-6}}{m}$$

$$m = \frac{0.24 \times 10^{-6}}{4.2 \times 10^{-6}} = 0.057\,kg$$

Equivalent dose

Absorbed dose only gives a rough guide to the amount of biological damage that can be done by the energy deposited in body tissue by radiation. In general, the higher the absorbed dose, the greater the biological effect. However, equal absorbed doses of radiation do not necessarily produce the same biological effects. For instance, $1.0\,Gy$ of alpha radiation is 20 times more harmful to a sample of tissue than $1.0\,Gy$ of gamma radiation. The reason for this is that alpha radiation causes a lot more ionisation in a short distance in a tissue and this causes severe damage to the cells affected. The ionisation produced by beta and gamma radiation is more spread out and the damage to the cells is less severe.

To take account of the damage done by different kinds of radiation on body tissue a quantity called the **radiation weighting factor** (w_R) is used. Radiation weighting factor does not have a unit – it is just a number. Values for the radiation weighting factor for different radiations are given on the *Data Sheet* on page 170.

The biological effect of the different types of radiation on a sample of body tissue is measured in terms of a quantity called **equivalent dose.**

$$\frac{\text{equivalent}}{\text{dose}} = \frac{\text{absorbed}}{\text{dose}} \times \frac{\text{radiation weighting}}{\text{factor}}$$

$$H = D \times w_R$$

where H = equivalent dose, measured in sieverts (Sv),
D = absorbed dose, measured in grays (Gy),
w_R = radiation weighting factor (a number with no unit).

This means that 1 sievert = 1 joule per kilogram
($1\,Sv = 1\,J\,kg^{-1}$).

Equivalent dose takes into account the type of the incident radiation (α, β, γ, slow or fast neutrons) and the energy of the incident radiation. This means that $1.0\,Sv$ of alpha radiation will have the same biological effect as $1.0\,Sv$ of beta radiation and $1.0\,Sv$ of gamma radiation. However, equivalent dose does not take into account the type of tissue exposed to the radiation.

Equivalent doses from different types of radiation, received by a sample of tissue, can be added together.

Worked examples

Example 1

A sample of tissue receives an absorbed dose of $4.6\,\mu Gy$ of slow neutrons. Calculate the equivalent dose received by the sample.

Solution

Note: $4.6\,\mu Gy = 4.6 \times 10^{-6}\,Gy = 0.000\,004\,6\,Gy$

$H = D \times w_R$ Note: w_R from *Data Sheet*

$H = 4.6 \times 10^{-6} \times 3 = 1.38 \times 10^{-5}\,Sv$

Example 2

A sample of tissue receives an equivalent dose of $50\,\mu Sv$ of alpha radiation. Calculate the absorbed dose received by the tissue.

Solution

Note: $50\,\mu Sv = 50 \times 10^{-6} = 0.000\,050\,Sv$

$H = D \times w_R$ Note: w_R from *Data Sheet*

$50 \times 10^{-6} = D \times 20$

$$D = \frac{50 \times 10^{-6}}{20} = 2.5 \times 10^{-6}\,Gy$$

Example 3

A technician working with a radioactive source receives an absorbed dose of $5.0\,\mu Gy$ of radiation. The equivalent dose received by the technician is $15\,\mu Sv$. Identify the radiation the technician was working with.

Solution

Note: $15\,\mu Sv = 15 \times 10^{-6}\,Sv = 0.000\,015\,Sv$
Note: $5\,\mu Gy = 5 \times 10^{-6}\,Gy = 0.000\,005\,Gy$

$H = D \times w_R$

$15 \times 10^{-6} = 5 \times 10^{-6} \times w_R$

$$w_R = \frac{15 \times 10^{-6}}{5 \times 10^{-6}} = 3$$

From *Data Sheet* an w_R of 3 is slow neutrons.

Example 4

A radiation worker estimates that she was irradiated by 2.5 mGy of slow neutrons and 50 μGy of fast neutrons. Calculate the total equivalent dose received by the worker.

Solution

Note: w_R values from *Data Sheet*

$$H_{slow} = D \times w_R = 2.5 \times 10^{-3} \times 3 = 7.5 \times 10^{-3}\,Sv$$

$$H_{fast} = D \times w_R = 50 \times 10^{-6} \times 10 = 500 \times 10^{-6}\,Sv$$

$$
\begin{aligned}
\text{Total equivalent dose } H &= H_{slow} + H_{fast} \\
&= 7.5 \times 10^{-3} + 500 \times 10^{-6} \\
&= 8 \times 10^{-3}\,Sv
\end{aligned}
$$

Equivalent dose rate

$$\text{equivalent dose rate} = \frac{\text{equivalent dose}}{\text{time taken}}$$

$$\dot{H} = \frac{H}{t}$$

where $\dot{H}$ = equivalent dose rate, measured in sieverts per second ($Sv\,s^{-1}$), sieverts per minute ($Sv\,min^{-1}$), sieverts per hour ($Sv\,h^{-1}$), etc.

H = equivalent dose, measured in sieverts (Sv),

t = time taken, measured in seconds (s), minutes (min), hours (h), etc.

Worked examples

Example 1

A radiation worker spends 4 hours in a radioactive area. The worker receives an equivalent dose of 80 μSv of radiation. Calculate the equivalent dose rate for the worker during this time.

Solution

Note: $80\,\mu Sv = 80 \times 10^{-6}\,Sv = 0.000\,080\,Sv$

$$\dot{H} = \frac{H}{t} = \frac{80 \times 10^{-6}}{4} = 2.0 \times 10^{-5}\,Sv\,h^{-1}$$

Example 2

A radiation worker receives an average equivalent dose rate of 15 μSv h⁻¹ during a 24-hour working week. Calculate the equivalent dose the worker receives in this time.

Solution

Note: $15\,\mu Sv\,h^{-1} = 15 \times 10^{-6}\,Sv\,h^{-1} = 0.000\,015\,Sv\,h^{-1}$

$$\dot{H} = \frac{H}{t}$$

$$15 \times 10^{-6} = \frac{H}{24}$$

$$H = 15 \times 10^{-6} \times 24 = 3.6 \times 10^{-4}\,Sv$$

Example 3

The cosmic-ray detector on board an aircraft indicates an average equivalent dose rate of 12 μSv h⁻¹ during a flight. A passenger, during the flight, receives an equivalent dose of 60 μSv. Calculate the time taken for the flight.

Solution

Note: $12\,\mu Sv\,h^{-1} = 12 \times 10^{-6}\,Sv\,h^{-1} = 0.000\,012\,Sv\,h^{-1}$

Note: $60\,\mu Sv = 60 \times 10^{-6}\,Sv = 0.000\,060\,Sv$

$$\dot{H} = \frac{H}{t}$$

$$12 \times 10^{-6} = \frac{60 \times 10^{-6}}{t}$$

$$t = \frac{60 \times 10^{-6}}{12 \times 10^{-6}} = 5.0\,h$$

Risk of biological harm to body tissue

The biological damage to body tissue and the total effect caused by exposure to radiation depends on:

- the absorbed dose, i.e. the energy absorbed per kilogram by the tissue
- the type of the ionising radiation – alpha, beta, gamma, fast neutrons, etc.
- the type of tissue exposed – some tissues, such as bone marrow or blood, are affected more by radiation than other parts of the body, such as muscles.

Background radiation

We are all exposed to both natural and man-made radiation – this is called **background radiation**.

When a Geiger–Müller tube and counter is left switched on in a room, radiation will be recorded on the counter – a number of counts – even though there is no radioactive source present.

By measuring the background counts over a period of time, the average number of counts recorded in a given time can be calculated – either the number of counts per second or the number of counts per minute. This is called the **background count rate**. When a radioactive source is placed near the counter, the measured count rate will be equal to the source count rate and the background count rate added together. The **source count rate** is obtained by subtracting the background count rate from the measured count rate. This gives the **corrected count rate**, i.e. the count rate from the radioactive source alone.

corrected count rate = total count – background
　　　　　　　　　　　　rate　　　　count rate

source count rate　·　= total count – background
　　　　　　　　　　　　rate　　　　count rate

Sources of radiation

The pie chart in Figure 7.7 shows the main sources of radiation that we are exposed to on Earth.

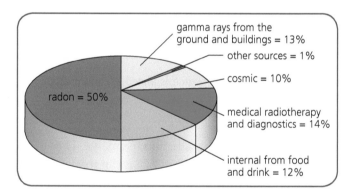

Figure 7.7 Sources of radiation dose

We are all exposed to natural background radiation. We breathe small amounts of the radioactive gas called radon. The ground and buildings around us are slightly radioactive. Our bodies contain some natural radioactivity from the food we eat and drink. We absorb cosmic rays from outer space. For the average person in the United Kingdom, the annual dose received from background radiation is about 2.2 mSv. However, individual doses vary greatly. For instance, if you take several aeroplane trips across the Atlantic each year, you will get a larger equivalent dose of cosmic ray radiation.

If you live in Aberdeen where the granite rocks are giving off radioactive radon gas, you are subjected to a much greater exposure to background radiation.

However, radiation can be beneficial to us, particularly in healthcare. We are exposed to radiation during medical examinations, such as chest and dental X-rays and the investigation of bone fractures and other diagnostic procedures. For these types of radiation the annual exposure for a member of the public should not exceed 1 mSv, while for radiation workers the annual exposure should not exceed 20 mSv. Radiation workers are permitted higher doses because:

- they are unlikely to be either old and infirm or young and vulnerable
- they will be subject to regular medical examination
- they will have their exposure monitored.

Handling radioactive sources

The radioactive warning sign (Figure 7.8) is displayed wherever radioactive sources are stored or are in regular use.

Figure 7.8 Radiation warning sign

When using radioactive sources, the following safety procedures should be followed:

- Always use tongs (see Figure 7.9) or wear special gloves when moving a source – never use bare hands.
- Arrange the source so that the radiation window points away from the body.
- Never point the source at your eyes.
- Wash your hands immediately after performing any experiment that involves radioactive sources.

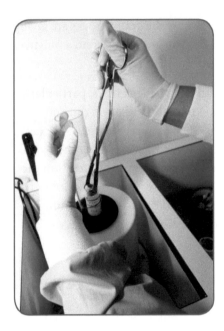

Figure 7.9 Using tongs to lift a radioactive source

Protection from radiation

There are three ways in which exposure to radiation can be reduced:

- Shielding – shield the source of radiation with an appropriate thickness of absorber.
- Limit the time of exposure – radioactive sources should be moved and used as quickly as possible to reduce the radiation present.
- Distance from source – the further you are from the source, the less radiation you will receive.

Using radiation in medicine

Treating cancer: radiotherapy

A cancerous tumour is a bundle of cells that reproduces in an uncontrolled manner. The treatment of cancer by radiation is called radiotherapy. The purpose of radiotherapy is to damage the cancer cells and stop them reproducing. As a result, the tumour then shrinks. However, during radiotherapy healthy cells are exposed to radiation and will be damaged. The damage to healthy tissue is minimised by calculating carefully the dose of radiation required to damage the cancer cells and by aiming the radiation at them as accurately as possible.

To reach the tumour inside the body, gamma radiation is used. The gamma source is held safely in a thick metal container. A slit in the container allows a controlled

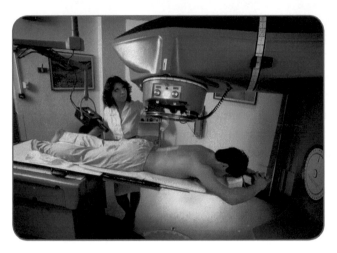

Figure 7.10 Radiotherapy

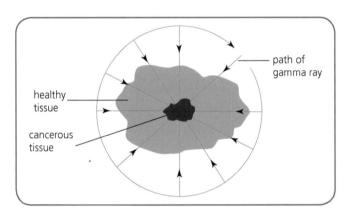

Figure 7.11 The gamma source is aimed at, and rotates around, the tumour

narrow beam of gamma radiation to emerge. The radiation source is arranged so that it can rotate around the patient. Rotating the source of radiation means that the tumour receives a large dose of radiation while the healthy tissue receives a much smaller dose.

Radiotherapy is more effective when given as a series of small doses.

Sterilisation

Gamma radiation can be used to kill bacteria and germs present on medical items such as syringes (sterilise them). The syringes are pre-packed and then exposed to an intense source of gamma radiation. This kills any bacteria and germs but does not make the syringe radioactive.

Radioactive tracers

Small amounts of radiation can be detected easily so the path of the radiation through an object can be followed. When a radioactive substance is used like this, it is called a **tracer**.

In medicine

The radioactive tracer is either injected into the patient or the patient drinks it. The substances used are chosen so that:

- they become concentrated in the organ that is to be examined
- they lose their radioactivity quickly
- they emit gamma radiation that can be detected outside the body (alpha and beta radiation would be absorbed inside the body).

A gamma camera can be used to detect the gamma radiation.

In industry

A leaking underground pipe can be detected by adding a radioactive tracer to the liquid in the pipe. Some of the liquid and tracer will leak into the soil surrounding the pipe. More gamma radiation will be detected at the surface at this point, thus indicating the position of the leak.

In agriculture

Tracers can be used to find out how well plants make use of fertilisers. To do this, a small amount of radioactive tracer is added to the fertiliser. The progress of the tracer through the plant can then be monitored.

Physics beyond the classroom

A common use of a tracer is to examine the working of a patient's kidneys. A radioactive form of an element called technetium is used as the tracer and is injected into the bloodstream. A gamma camera is used to detect the gamma rays from the patient's body.

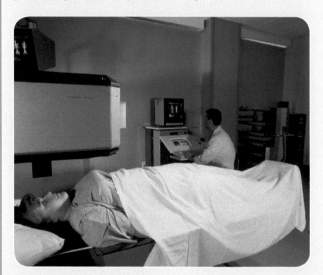

Figure 7.12 A gamma camera

The radioactive material is filtered out of the bloodstream by the kidneys and within a few minutes the radiation is concentrated in the kidneys. After about 10 minutes all of the radiation should be transferred to the bladder. The gamma camera is used to take images every few seconds for about 20 minutes. A computer connected to the gamma camera builds up a picture of the amount of radiation in each kidney; this can be in the form of a graph.

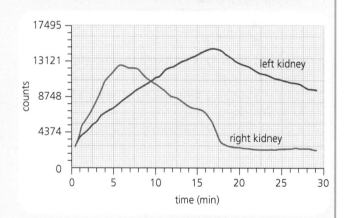

Figure 7.13 Graph showing how the amount of radiation during a kidney examination varies with time

The graph shows that the amount of the radiation in the right kidney falls more quickly than that from the left kidney. This indicates that the left kidney is not working properly.

Key facts and physics equations: effects of nuclear radiation

- A simple model of the atom includes protons, neutrons and electrons.
- An alpha (α) particle is a helium nucleus (two protons and two neutrons) and it is positively charged.
- A beta (β) particle is a high-speed electron and it is negatively charged.
- A gamma (γ) ray is an electromagnetic wave and it travels at $3 \times 10^8 \, \mathrm{m\,s^{-1}}$ in air. It is a burst of energy and a transverse wave.
- The energy from radiation is absorbed in the medium through which it passes.
- An alpha particle is absorbed by a few centimetres of air or a thin sheet of paper.
- A beta particle is absorbed by a few metres of air or a few millimetres of aluminium.
- A gamma ray is absorbed by a few kilometres of air or a few centimetres of lead.
- Ionisation is the breaking up of a neutral atom into positive and negative pieces.
- Alpha particles are more ionising than beta particles or gamma rays.
- Absorbed dose is the energy absorbed per unit mass of the absorbing material.
- Absorbed dose = $\dfrac{\text{energy absorbed}}{\text{mass of absorbing material}}$

  i.e $\qquad D = \dfrac{E}{m}$
- Absorbed dose, D, is measured in grays (Gy), energy absorbed, E, is measured in joules (J) and mass of absorbing material, m, is measured in kilograms (kg).
- 1 gray = 1 joule per kilogram ($1\,\mathrm{Gy} = 1\,\mathrm{J\,kg^{-1}}$)

- A radiation weighting factor is given to each kind of radiation as a measure of its biological effect.
- Equivalent dose = absorbed dose × radiation weighting factor, i.e. $H = D \times w_R$
- Equivalent dose, H, is measured in sieverts (Sv), absorbed dose, D, is measured in grays (Gy) and radiation weighting factor, w_R, is a number and has no unit.
- 1 sievert = 1 joule per kilogram ($1\,\mathrm{Sv} = 1\,\mathrm{J\,kg^{-1}}$)
- Equivalent dose rate = $\dfrac{\text{equivalent dose}}{\text{time taken}}$

  i.e. $\qquad \dot{H} = \dfrac{H}{t}$
- Equivalent dose rate, $\dot{H}$, is measured in sieverts per second ($\mathrm{Sv\,s^{-1}}$), sieverts per hour ($\mathrm{Sv\,h^{-1}}$) etc., equivalent dose, H, is measured in sieverts (Sv) and time taken, t, is measured in seconds (s), hours (h) etc.
- For a member of the public, average annual background radiation is 2.2 mSv.
- For a member of the public, annual dose due to medical radiation should not exceed 1 mSv.
- For a radiation worker, annual dose should not exceed 20 mSv.
- The risk of biological harm from an exposure to radiation depends on the absorbed dose, the type of radiation, e.g. α, β, γ, slow neutron, and the tissue or body organs exposed.
- Equivalent dose can be reduced by shielding, by limiting the time of exposure and by increasing the distance from a source.
- When handling radioactive sources always use tongs, point the source window away from the body, never point the source at your eyes and wash hands immediately afterwards.

End-of-chapter questions

Information, if required, for use in the following questions can be found on the *Data Sheet* on page 170.

1 A simple model of the atom contains protons, neutrons and electrons. Draw a diagram of the atom. Label and name each of the particles and state the nature of their charges.

2 A radioactive source emits alpha particles and beta particles.
   a) What is an alpha particle? State the sign of its charge.
   b) What is a beta particle? State the sign of its charge.

3 Alpha, beta and gamma radiations all produce ionisation.
   a) What is meant by the term *ionisation*?
   b) In a given thickness of material, which of the above radiations is most likely to produce the greatest amount of ionisation?
   c) Which of the above radiations has the greatest range in air?

4 A technician working with radioactivity wears a film badge to monitor the amount of radiation she is exposed to. The film is contained in a plastic holder with windows of different materials as shown in Figure 7.14. No light can reach the film.

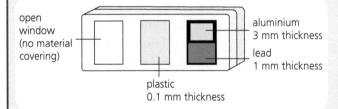

Figure 7.14

   a) Why does the technician's level of exposure to radiation have to be carefully monitored?
   b) The technician is exposed to beta radiation. Which window or windows on the diagram will be affected by the beta radiation? Give a reason for your choice(s).

5 A hospital uses a radioactive tracer to investigate the blood flow in a patient's kidney. The radiation from the tracer is detected outside the patient's body. Sources of alpha, beta or gamma radiation are available to use as the tracer. Which source of radiation would be suitable to use as the tracer in this investigation? Explain your answer.

6 A technician is asked to check that sources of alpha, beta and gamma radiation are labelled correctly. The technician has the following absorber materials:
   ● a thin sheet of paper
   ● 5 mm thick aluminium
   ● 20 mm thickness of lead
   Describe how the technician would identify each of the sources.

7 State the three factors that affect the risk of biological harm from radiation.

8 The mass of a sample of tissue is 0.25 kg. The tissue is exposed to radiation and absorbs 15 μJ of energy. Calculate the absorbed dose received by the tissue.

9 The mass of a sample of tissue is 20 g. The tissue receives an absorbed dose of 8.5 mGy from a source of alpha radiation.
   a) What is meant by the term *absorbed dose*?
   b) Calculate the energy absorbed by the tissue.
   c) Calculate the dose equivalent received by the tissue.

10 Each type of radiation has a radiation weighting factor.
   a) What information does the radiation weighting factor give about the radiation?
   b) What is the radiation weighting factor for fast neutrons?

11 The mass of a sample of tissue is 125 g. The sample is exposed to gamma radiation and absorbs 4.5 mJ of energy. Calculate the equivalent dose received by the tissue.

12 A sample of tissue receives an absorbed dose of 75 μGy of alpha radiation and 15 mGy of slow neutrons.
   a) Calculate the total equivalent dose received by the tissue.
   b) Describe two ways in which the absorbed dose received by the tissue could be reduced.

13 A sample of tissue is exposed to alpha particles. The tissue receives an equivalent dose of 200 μSv. Calculate the absorbed dose received by the tissue.

14 A sample of tissue is exposed to a source of radiation. The tissue receives an absorbed dose of 117 μGy and an equivalent dose of 1.17 mSv.
   a) Calculate the radiation weighting factor for the radiation.
   b) Identify the radiation used.

15 A radiation worker spends 8 hours in an area where the equivalent dose rate is 210 μSv h^{-1}. Calculate the equivalent dose received by the worker in the 8 hours.

8   Using the nucleus

Activity of a source

A radioactive material contains atoms, the nuclei of which are unstable. The radioactive nuclei may become stable by emitting alpha, beta or gamma radiation. When a nucleus emits alpha, beta or gamma radiation the nucleus is said to disintegrate or **decay**. The number of radioactive nuclei that decay in 1 second is called the **activity** of the radioactive source.

$$\text{activity} = \frac{\text{number of radioactive nuclei decaying}}{\text{time taken}}$$

$$A = \frac{N}{t}$$

where A = activity of source, measured in becquerels (Bq), N = number of radioactive nuclei decaying (this is a number and does not have a unit), t = time, measured in seconds (s).

- One becquerel means that one radioactive nucleus has decayed in 1 second ($1\,\text{Bq} = 1\ \text{decay}\,\text{s}^{-1}$).
- The activity of a radioactive source cannot be measured experimentally.
- The activity of radioactive sources is usually measured in kilobecquerels (kBq) or megabecquerels (MBq).
  ($1\,\text{kBq} = 10^3\,\text{Bq} = 1000\,\text{Bq};$
  $1\,\text{MBq} = 10^6\,\text{Bq} = 1\,000\,000\,\text{Bq}$)

Worked examples

Example 1

In a radioactive source, 200 000 nuclei decay in 90 s. Calculate the activity of the source.

Solution

$$A = \frac{N}{t} = \frac{200\,000}{90} = 2222\,\text{Bq}$$

(Do not put too many figures in your answer. This answer is better written as $2.2 \times 10^3\,\text{Bq}$.)

Example 2

A radioactive source has an activity of 80 MBq. Calculate the number of nuclei that decay in the source in 40 s.

Solution

Note: 80 MBq = 80×10^6 Bq = 80 000 000 Bq

$$A = \frac{N}{t}$$

$$80 \times 10^6 = \frac{N}{40}$$

$$N = 80 \times 10^6 \times 40 = 3.2 \times 10^9 \text{ decays}$$

Example 3

The activity of a radioactive source is 20 000 Bq. Calculate the time taken for 130 000 disintegrations to occur in the radioactive source.

Solution

$$A = \frac{N}{t}$$

$$20\,000 = \frac{130\,000}{t}$$

$$t = \frac{130\,000}{20\,000} = 6.5 \text{ s}$$

Half-life

The activity of any radioactive source decreases with time (since the number of radioactive nuclei present is decreasing as the nuclei decay). The graph in Figure 8.1 shows how the activity of a radioactive source varies with time.

Any radioactive source gives a graph with a similar shape.

When a radioactive substance disintegrates, the activity (the number of radioactive nuclei that disintegrate in 1 second) depends only on the number of radioactive nuclei present.

- The half-life of a radioactive source is the time taken for one-half of the radioactive nuclei to disintegrate.
- The half-life of the radioactive source is also the time taken for the activity of the radioactive source to halve.
- The half-life of a radioactive source is unique and can be used to identify a radioactive source (similar to a fingerprint).
- Half-lives can vary from a fraction of a second to thousands of years as shown in Table 8.1.

Source	Half-life
Uranium-238	4 500 000 000 years
Radium-226	1600 years
Strontium-90	28 years
Cobalt-60	5.3 years
Sodium-24	15 hours
Copper-66	5.2 minutes
Bismuth-212	60.5 seconds

Table 8.1

The activity of a radioactive source is not affected by shielding or distance from the source.

In practice, the count rate recorded from a source is a measure of the activity of the source.

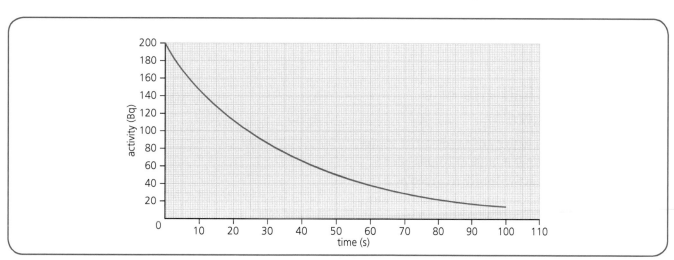

Figure 8.1 A graph of activity of a radioactive source against time

Worked examples

Example 1

The activity of a radioactive source is 1600 MBq. The activity of the source 40 days later is 50 MBq. Calculate the half-life of the source.

Solution

The half-life is the time taken for the activity of the source to halve.

Number of half-lives	Activity (MBq)	Time (days)
0	1600	0
1	800	-
2	400	-
3	200	-
4	100	-
5	50	40

5 half-lives = 40 days

1 half-life = $\frac{40}{5}$ = 8 days

Example 2

A radioactive source has an activity of 160 kBq. The half-life of the source is 6.0 hours. Calculate the time taken for the activity to fall to 10 kBq.

Solution

The half-life is 6 hours so the activity will halve every 6 hours.

Number of half-lives	Activity (kBq)	Time (h)
0	160	0
1	80	6
2	40	12
3	20	18
4	10	24

Time taken for the activity to fall to 10 kBq is 24 hours.

Example 3

A radioactive source has a half-life of 8 days. At the start of an experiment the total count rate recorded is 1225 counts per minute (cpm). Find the total recorded count rate after 24 days. The background count rate on both occasions is 25 counts per minute.

Solution

source count rate = total count rate − background count rate
= 1225 − 25 = 1200 cpm

Number of half-lives	Source count rate (cpm)	Time (days)
0	1200	0
1	600	8
2	300	16
3	150	24

At 24 days, source count rate = 150 cpm
total count rate = source count rate + background count rate
= 150 + 25 = 175 cpm

Measuring the half-life of a radioactive source

We cannot measure the activity of a radioactive source. The count rate (counts per minute or counts per second) from the radioactive source recorded by a Geiger–Müller tube and counter is taken as a measure of the activity of the source (Figure 8.2).

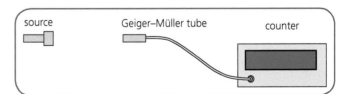

Figure 8.2 A Geiger–Müller tube and counter being used to measure count rate

- Measure the background count rate in counts per minute (cpm) (in the absence of the radioactive source).
- Place the radioactive source a fixed distance in front of the Geiger–Müller tube and measure the total count rate (this is the total count rate at $t = 0$).
- Measure the total count rate at regular time intervals.
- Correct all of the total count rates for background radiation to find the source count rate:

$$\frac{\text{source}}{\text{count rate}} = \frac{\text{total}}{\text{count rate}} - \frac{\text{background}}{\text{count rate}}$$

- Draw a graph of source count rate against time.
- Using the graph, work out the half-life (the time taken for the source count rate to halve).

Generating electricity: power stations

Electrical energy is generated mainly in large, central power stations fuelled by coal, gas or nuclear fuel. These power stations operate in similar ways:

- Water is heated to form steam, which turns a turbine.
- The turbine drives a generator to produce electrical energy.
- The exhaust steam from the turbine is condensed to water and returned to the boiler or reactor to be reheated.

Normally fossil fuels (coal and gas) are burnt to release the heat needed to boil the water (how heat is released from nuclear fuel is discussed later). However, the burning of fossil fuels has a number of important disadvantages:

- There is a finite amount of fossil fuels available – they will eventually run out.
- When they are burnt it causes air pollution.
- They are not renewable.

There are a number of alternative ways of generating electricity:

- hydroelectric power
- wind power
- solar power
- tidal power
- geothermal power – energy from the Earth's core
- biomass – using plants.

Nuclear energy

Advantages of using nuclear energy include:

- No gases are produced to cause air pollution.
- Although there are limited resources of uranium – the fuel used in a nuclear power station – there are sufficient quantities to provide large amounts of fuel.
- Small amounts of uranium – several kilograms – can supply several power stations.
- For the same mass of fuel, uranium produces more energy than the fossil fuels.

The main disadvantages of using nuclear fuel are:

- the disposal of the nuclear waste produced
- the possibility of an accident – a nuclear explosion
- limited resources of uranium.

Nuclear fission

Nuclear reactions normally take place inside a nuclear reactor. These reactions can cause a lot of heat to be produced.

Neutrons are found inside the nucleus of an atom. When a neutron is fired into the nucleus of a uranium atom, the nucleus splits into two pieces, called fragments, and two or more neutrons are released. This process is called **nuclear fission** (see Figure 8.3). In nuclear fission a heavy nucleus splits into two lighter nuclei with the release of two or three neutrons.

In a nuclear power station a material called uranium-235 (U-235) is used as the fuel.

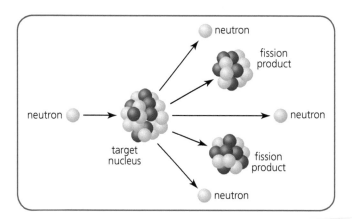

Figure 8.3 Nuclear fission

During the process of fission, nuclear potential energy is transferred to the fission fragments as kinetic energy.

95

The kinetic energy of the fission fragments causes heat to be produced.

The neutrons released during the fission process can cause further fissions. This in turn can cause even more fissions and so on. This is called a **chain reaction** (see Figure 8.4). The mass of material needed to produce a chain reaction is called the **critical mass**.

An uncontrolled chain reaction is the reaction that takes place in a nuclear bomb. In a nuclear power station, the chain reaction must be controlled so that:

- sufficient power is produced for our needs
- the reactor does not explode.

The amount of energy produced per reaction is very small, but there is a tremendous amount of energy produced in a nuclear reactor where vast numbers of reactions take place each second.

When nuclei undergo fission the energy produced is given to the fission products as kinetic energy (they leave the reaction with very high speeds). This kinetic energy is eventually turned into heat as the fission products collide with other atoms.

The fission reaction results in a decrease in mass. This decrease in mass means that energy is released.

Nuclear fusion

Nuclear fusion involves two light nuclei joining together to form a heavier nucleus (Figure 8.5). During the fusion reaction energy is released. Nuclear fusion occurs in the hydrogen bomb. In the Sun and other stars nuclear fusion produces vast quantities of energy.

Most work on fusion involves two hydrogen nuclei called deuterium and tritium. These particles are both positively charged and as a result repel one another (like charges repel). The repulsion can be overcome by heating the particles to about 100 million degrees Celsius and placing them under very high pressure; the very hot ionised gas is called **plasma**. When the particles collide at these high temperatures and pressures the particles fuse together.

The fusion reaction results in a decrease in mass. This decrease in mass means that energy is released. This energy is carried away as kinetic energy of the products of the fusion reaction.

If 1 gram of deuterium and tritium were fused in this way, the energy released would be enough to supply an average household for 40 years. There are, however, technical difficulties that scientists are working hard to overcome.

In the hydrogen bomb, these nuclei are speeded up by raising their temperature to over a hundred million degrees Celsius. To produce such a high temperature a fission bomb is needed. This is included in a hydrogen bomb. When a collision occurs between the fast-moving nuclei at these high temperatures, the reaction is called **thermonuclear fusion**.

In the process of fission and fusion the energy produced has not been 'created'. It was present as 'potential energy' in the nucleus or nuclei before fission or fusion took place. The fission or fusion process caused the energy to be released after the reaction had taken place.

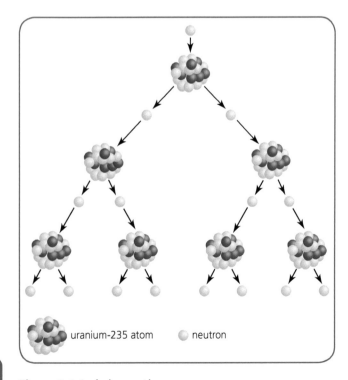

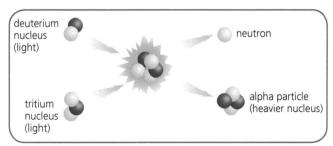

Figure 8.4 A chain reaction

Figure 8.5 A fusion reaction

The nuclear reactor

In a nuclear power station, a nuclear reactor is used to generate the heat needed to change water into steam.

The five main parts of a reactor are:

- **Fuel** – this is in the form of rods filled with pellets of uranium.
- **Moderator** – during fission the neutrons released are very fast moving. Further fission is more likely if the neutrons are slowed down using a moderator. The moderator is usually carbon in the form of graphite (water can also be used). The moderator does not absorb neutrons but slows them down as the neutrons collide with the graphite atoms.
- **Control rods** – the amount of electrical energy needed varies, with greater demand during the day than through the night. This means that – apart from safety reasons – the number of fission reactions have to be controlled. This is done using control rods made from boron or cadmium. The control rods absorb neutrons. When the rods are lowered into the reactor, the number of fission reactions decreases. This regulates the heat output from the core of the reactor. The rods are raised and lowered automatically in response to changes in temperature in the core. In an emergency all the control rods are inserted into the reactor core. The fission reactions are shut down completely.
- **Coolant** – the heat produced by the fission reactions is removed by pumping coolant past the hot fuel elements. The hot fluid – liquid or gas – is piped from the core to a heat exchanger (pipes). The heat passes through the pipes and is absorbed by water, which changes into steam. This steam is used to drive turbines. Since the steam is not radioactive it can be cooled down and reused. The coolant, having transferred heat to the water, is returned to the core to be reheated.
- **Containment vessel** – the core of the nuclear reactor is contained under a thick shield made of concrete reinforced with steel. This prevents the escape of radioactive materials and neutrons. The vessel is also able to withstand the high temperatures and pressures of the reactor when in normal use.

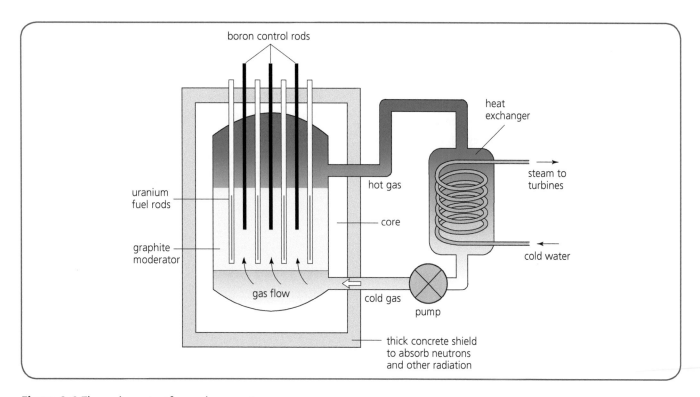

Figure 8.6 The main parts of a nuclear reactor

Nuclear waste

One of the major problems of using nuclear energy is the radioactive waste produced. The three categories of waste product are:

- low-level radioactive solid waste
- medium-level radioactive waste
- high-level radioactive waste.

Low- and medium-level radioactive waste does not mean that it is not dangerous. It means that the radioactivity is less concentrated but the waste still has to be stored.

High-level radioactive waste is mainly spent nuclear fuel. The materials left in the fuel rods are highly radioactive and have long half-lives. This means that the waste has to be stored for a very long time in a suitable environment before the radiation reaches a safe level.

The spent fuel rods are stored initially in cooling ponds. The water removes some of the heat and also absorbs the radiation. The fuel rods are then removed and sealed in drums. The drums are then stored in vaults.

Nuclear accidents

Despite strict safety precautions, accidents involving nuclear energy are possible. This is often due to human error.

In 1986, a nuclear disaster took place at Chernobyl in the Ukraine. While carrying out tests at the core of the nuclear reactor, technicians had turned off vital safety systems. This led to an uncontrolled chain reaction and an explosion that blew a hole in the reactor. Large amounts of radiation were released into the atmosphere. The radiation was carried across Europe by the wind and was detected in Britain.

Key facts and physics equations: using the nucleus

- The activity of a radioactive source is the number of nuclei decaying in 1 second.
- Activity = $\dfrac{\text{number of radioactive nuclei decaying}}{\text{time taken}}$

  i.e. $A = \dfrac{N}{t}$
- Activity of source, A, is measured in becquerels (Bq), number of radioactive nuclei decaying, N, is a number and does not have a unit and time, t, is measured in seconds (s).

- One becquerel is one radioactive nucleus decaying in 1 second.
- The activity of a radioactive source decreases with time.
- The half-life of a radioactive source is the time taken for half of the radioactive nuclei present to disintegrate.
- Fission involves a heavy nucleus splitting into two lighter nuclei with the release of two or three neutrons.
- Fusion involves two light nuclei joining together to form a heavier nucleus.

End-of-chapter questions

Information, if required, for use in the following questions can be found on the *Data Sheet* on page 170.

1 A radioactive source has an activity of 3.5 MBq.
   a) Explain what is meant by *an activity of 3.5 MBq*.
   b) State **two** safety precautions that would be taken when handling this radioactive source.

2 A radioactive source has a half-life of 5.3 years.
   a) Explain what is meant by the statement 'half-life of 5.3 years'?
   b) Calculate how many years it takes for the activity of the radioactive source to decrease to 1/16th of its original value.

3 At the start of a medical examination a radioactive source has an activity of 1600 kBq. The half-life of the source is 6 hours. Calculate the activity of the source after 18 hours.

4 The age of a prehistoric wooden axe-handle can be determined from the activity of radioactive carbon-14 in the wood. The half-life of carbon-14 is 5700 years. When the axe-handle was carved, the activity of carbon-14 in the wood was 2400 µBq. At the present time, the activity of carbon-14 in the wood is 150 µBq. Calculate the age of the axe-handle.

5 Figure 8.7 shows how the corrected count rate from a radioactive source varies with time. From the graph, determine the half-life of the source.

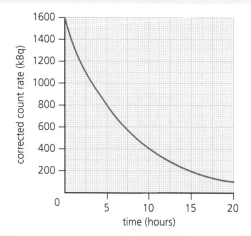

Figure 8.7

6 What is meant by the term *nuclear fusion*?

Exam practice for Chapters 6–8

Information, if required, for use in the following questions can be found on the *Data Sheet* on page 170.

1 a) There are two types of wave, longitudinal and transverse.
      i) State an example of a longitudinal wave.
      ii) What is transferred by longitudinal and transverse waves?
   b) The behaviour of water waves is observed in a long tank.

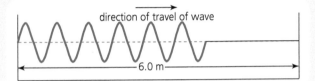

Figure E2.1

Water waves are created by a wave machine at one end of the tank. The machine produces 20 waves every minute. The length of the water tank is 6.0 m. A wave crest takes 12 s to travel the length of the tank.
      i) Calculate the speed of the water waves.
      ii) Calculate the wavelength of the water waves.

2 a) The electromagnetic spectrum is shown below in order of increasing frequency.

Radio waves	Television waves	P	Infrared	Visible light	Q	X-rays	Gamma rays

      i) Name the missing radiations P and Q.
      ii) Name a detector for X-rays.
   b) A source of microwaves can transmit on the following three frequencies:
      1.5×10^{11} Hz 2.5×10^{11} Hz 1.8×10^{10} Hz
      Calculate the wavelength of the microwaves with the longest wavelength.

3 A ray of red light is incident on a rectangular glass block as shown in Figure E2.2.

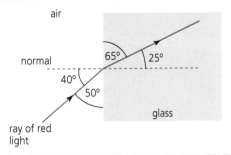

Figure E2.2

a) i) State the size of the angle of incidence.
   ii) State the size of the angle of refraction.
b) In air, the frequency of the red light is 4.8×10^{14} Hz.
   i) Calculate the wavelength of the red light in air.
   ii) What is the frequency of the red light in the glass block?

4 A student studies the behaviour of water waves in a water tank. She notices that the waves undergo diffraction.

a) What is meant by the term *diffraction*?
b) Copy and complete the diagram below to show the waves after they have passed the obstacle.

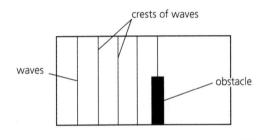

Figure E2.3

c) The frequency of the waves in the tank is increased.
   Will this change increase, decrease or keep the amount of diffraction the same?
   You must explain your answer.

5 At a paper mill, the thickness of cardboard is measured using a radioactive source, a Geiger–Müller tube and a counter.

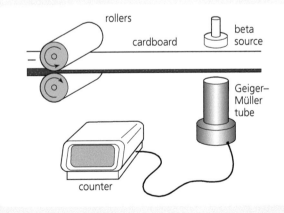

Figure E2.4

a) A source of beta radiation is used as the radioactive source. Explain why a source of alpha or gamma radiation would be **unsuitable** to measure the thickness of the cardboard.

b) What happens to the recorded count rate as the thickness of the cardboard decreases?

c) State **two** safety precautions that should be taken when handling radioactive sources.

6 Part of a smoke detector is shown in Figure E2.5.

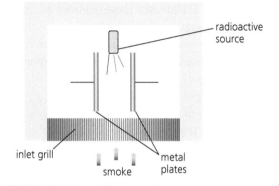

Figure E2.5

The radioactive source causes ionisation of the air between the two metal plates. This produces a small current in an electrical circuit in the smoke detector. When there are smoke particles in the air between the two metal plates the current is reduced and the smoke alarm sounds.

a) State what is meant by the term *ionisation*?

b) For the smoke detector to work the radioactive source has to produce a lot of ionisation of the air. A source of alpha, or beta or gamma radiation is available for use in the smoke detector. Explain in terms of ionisation which radioactive source should be used.

7 A radioactive source is to be used in the internal examination of a patient's kidney. The procedure involves taking measurements of the radioactivity present in the kidney for 30 minutes. Three different radioactive sources are available for the procedure:

- Source X – a gamma source with a half-life of 6.0 hours.
- Source Y – a beta source with a half-life of 6.7 hours.
- Source Z – a gamma source with a half-life of 4.3 days.

a) Which source should be selected for the kidney investigation?

b) Explain why the other two sources are not suitable for the kidney investigation.

c) Radioactive materials are sometimes injected into a patient to obtain images of a part of the patient's body. Give **two** reasons why alpha-emitting materials would be unsuitable to inject into a patient's body for this purpose.

8 Technetium is a source of gamma radiation. To investigate the condition of a patient's lungs, the patient is injected with a solution of technetium.

a) Explain why a source of gamma radiation must be used for this investigation.

b) The gamma radiation is detected by a gamma camera. When the technetium solution is injected into the patient it has an activity of 640 MBq. The half-life of the technetium is 6.0 hours.

i) State what is meant by the term *half-life*.

ii) Calculate the activity of the solution after 24 hours.

c) The effect that a dose of radiation has on living material depends on a number of factors. State **two** of these factors.

9   A radioactive source emits alpha particles. An experiment is carried out to determine the half-life of the radioactive source. The graph shows how the total count rate recorded in the experiment varies with time.

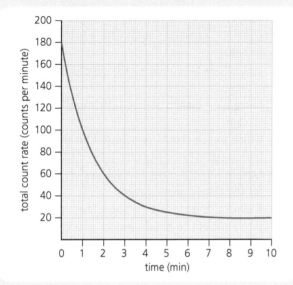

Figure E2.6

a) What is meant by *alpha particles*?
b) Estimate the average background count rate in counts per minute.
c) Determine the half-life of the radioactive source.

10 At a power station, a nuclear reactor is used in the generation of electricity. In the reactor, uranium nuclei are bombarded by neutrons. Some of the uranium nuclei undergo fission.
a) What is meant by the term *fission*?
b) In a fission reaction, two or three neutrons are released. Explain why these neutrons are important in the operation of the reactor.
c) An engineer working in the power station is exposed to the following absorbed doses of radiation:
   • 3.0 µGy of fast neutrons
   • 2.0 µGy of slow neutrons
   • 1.5 µGy of gamma rays.
   Calculate the total equivalent dose received by the engineer.

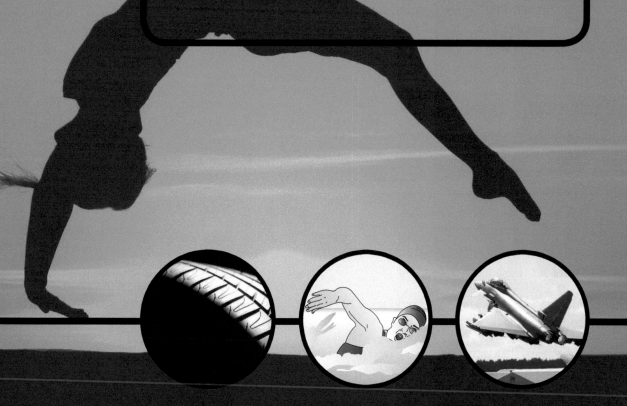

Section 5

Dynamics

9   Kinematics

Average speed

Consider David running a race between points A and B as shown in Figure 9.1.

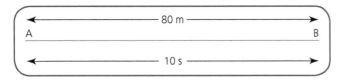

Figure 9.1 Measuring average speed

Points A to B are 80 m apart. David takes 10 s to run from A to B.

This means that David travels, on average, 8 m every second – an average speed of 8 metres per second (8 m s^{-1}).

The average speed is the steady (constant) speed David must travel, all of the time, to cover 80 m in a time of 10 s.

To measure average speed requires the measurement of the distance travelled and the time taken to travel this distance.

$$\text{average speed} = \frac{\text{distance travelled}}{\text{time taken}}$$

$$\text{average speed} = \frac{d}{t}$$

where average speed is measured in metres per second (m s^{-1}),
d = distance travelled, measured in metres (m),
t = time taken, measured in seconds (s).

Sometimes average speed is given the symbol $\bar{v}$.

Worked examples

Example 1

A cyclist takes 5 minutes to travel a distance of 1500 m. Calculate the average speed of the cyclist.

Solution

Note: 5 minutes = (5 × 60) s

$$\text{average speed} = \frac{d}{t} = \frac{1500}{(5 \times 60)} = 5.0\,\text{m}\,\text{s}^{-1}$$

Example 2

A car travels a distance of 12 km at an average speed of 20 m s^{-1}. Calculate the time taken for the car to travel the 12 km.

Solution

Note: 12 km = 12 × 10^3 m = 12 000 m

$$\text{average speed} = \frac{d}{t}$$

$$20 = \frac{12 \times 10^3}{t}$$

$$t = \frac{12 \times 10^3}{20} = 600\,\text{s}$$

Example 3

During a journey between two stations a train travels at an average speed of 40 m s^{-1}. The train takes 1.5 hours to make the journey. Calculate the distance travelled by the train.

Solution

Note: 1.5 h = (1.5 × 60) minutes = (1.5 × 60 × 60) s

$$\text{average speed} = \frac{d}{t}$$

$$40 = \frac{d}{(1.5 \times 60 \times 60)}$$

$$d = 40 \times (1.5 \times 60 \times 60) = 216\,000\,\text{m}$$

Average speed and instantaneous speed

During David's 80 m race, his speed will not always be constant at 8.0 m s^{-1}, but will change. At the start of the race he is stationary and so his speed is zero. When he starts, his speed will increase rapidly until he reaches his maximum speed. Towards the end of the race, as David begins to tire, his speed will decrease slightly. This means that the average speed and the speed at any particular time – the **instantaneous speed** – during the race are unlikely to be the same.

Instantaneous speed is the speed of an object at a particular time (or instant).

What if we wish to measure David's speed at, say, point X during the race?

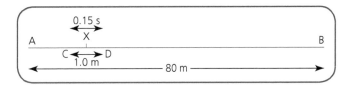

Figure 9.2 Measuring instantaneous speed

Since the time interval between A and B is large, David's speed at X is unlikely to be the same as the average speed for the whole race. However, by measuring the time taken for David to travel from point C (just before X) to point D (just after X) gives:

distance CD = 1.0 m

time to travel from C to D = 0.15 s

$$\text{average speed between C and D} = \frac{1.0}{0.15} = 6.7\,\text{m}\,\text{s}^{-1}$$

This is a better estimate of the speed at X since the time interval involved is very small. This means that any change in speed as David moves between C and D will be very small.

The instantaneous speed of an object is approximately the same as the average speed of the object as long as the time used in the equation for average speed is very small.

Measuring instantaneous speed

To measure short time intervals requires the use of a light gate connected to an electronic timer. Figure 9.3 shows a trolley moving down a sloping track.

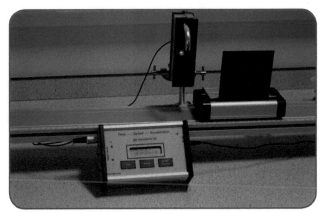

Figure 9.3 Measuring instantaneous speed using a light gate and a timer

When a card mounted on the trolley breaks the light beam of the light gate, the timer starts timing. When the card has passed through the light gate, the light beam is restored and the timer stops timing. The time taken for the card to pass through the light gate is recorded on the timer.

$$\text{instantaneous speed} = \frac{\text{length of card}}{\text{time on timer}}$$

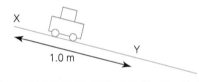

Worked example

Example

A card of length 100 mm is attached to a vehicle. The vehicle is released from point X and runs down a slope.

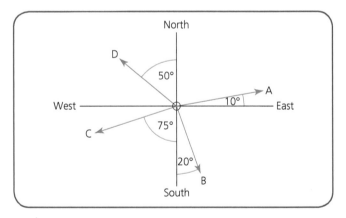

Figure 9.4

The card passes through a light gate positioned at point Y.

distance from X to Y = 1.0 m

length of card that passes through light gate = 100 mm

time for trolley to travel from X to Y = 2.50 s

time for card to pass through light gate = 0.125 s

a) Calculate the average speed of the trolley between X and Y.

b) Calculate the instantaneous speed of the trolley at Y.

Solution

Note: 100 mm = 100 × 10^{-3} m = 0.100 m

a) average speed $= \dfrac{d}{t} = \dfrac{1.0}{2.5} = 0.4\,\text{m s}^{-1}$

b) instantaneous speed at Y $= \dfrac{d}{t} = \dfrac{100 \times 10^{-3}}{0.125} = 0.8\,\text{m s}^{-1}$

The speed at Y is approximately the instantaneous speed as the time for the card to pass through the light gate is very small.

Scalars and vectors

A **scalar quantity** is a quantity that requires only a size (how large it is) and a unit, e.g. 10 m, 5.2 kg, 12 s.

A **vector quantity** is a quantity that requires both a size (and a unit) and a direction, e.g. 10 m east, 5 m s^{-1} south. When stating a vector quantity the size, the unit and the direction are required.

There are four vector quantities in this course (all other quantities are scalar quantities). They are:

- displacement
- velocity
- acceleration
- force.

Note that weight is the name given to a particular force (see Chapter 10). This means that weight is also a vector.

Using compass points to describe direction

A vector quantity can be represented by a straight line. An arrow must be placed on the line to show the direction of the vector quantity. Four vectors are shown in Figure 9.5.

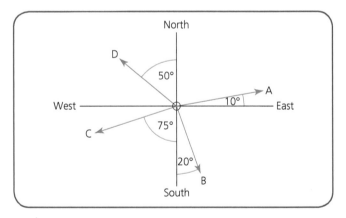

Figure 9.5 Vector examples

Vector OA is 10° north of the line labelled east. This is shortened to 10° N of E.

Vectors can also be described using bearings. A bearing is written as a three-digit number without the degree symbol. North has a bearing of 000. East is 90° clockwise from north and has a bearing of 090, south is 180 and west is 270. So vector OA has a bearing of 080 (90° − 10°, written as 080). The directions of the four vectors are shown in Table 9.1.

Vector	Direction	Bearing
OA	10° N of E	080
OB	20° E of S	160
OC	75° W of S	255
OD	50° W of N	310

Table 9.1

Two vectors can only be equal if they have the same size and the same direction.

Distance and displacement

Distance is a scalar quantity. It is how far the object travels.

Displacement is a vector quantity. It is the distance travelled in a particular direction by an object. It is the shortest distance between the start and finish of a journey – the straight line joining the start to the finish.

Where two or more displacements are added together, the resulting displacement (from the starting point to the finishing point – from the 'tail' of the first vector to the 'head' of the last vector) is known as the **resultant displacement**.

When dealing with vectors, always draw a diagram showing the size and direction of each vector. This is called a **vector diagram**.

Worked examples

Example 1

A girl walks 200 m east. She then turns and walks 50 m west. Calculate:
a) the distance travelled by the girl
b) the resultant displacement of the girl.

Solution

a) Distance travelled = 200 + 50 = 250 m

b) Figure 9.6 shows her walk as a scale vector diagram (a vector diagram can be used for any kind of vector). In this case, a 1 cm line represents 25 m of the girl's walk. Use a ruler to check that each displacement of the girl has the correct length and that the resultant displacement (the result of the two displacements – from the 'tail' of the first vector to the 'head' of the last vector) has the correct length. To distinguish it from the other vectors, the resultant displacement is normally shown with a double arrow.

For accuracy, scale vector diagrams should be as large as possible.

When the vectors act along the same straight line, positive and negative signs are used to indicate the direction of the vector. A vector going in one direction is positive; a vector going in the opposite direction is negative.

In this example, taking a vector going east as positive gives 200 m east = (+)200 m and a vector going 50 m west is –50 m. So adding the two vectors gives:

    resultant displacement = 200 + (–50) = (+)150 m

Since this is positive, it means 150 m east.

    resultant displacement = 150 m east

Note: when using a scale vector diagram, the resultant vector is drawn from the 'tail' of the first vector to the 'head' of the last vector.

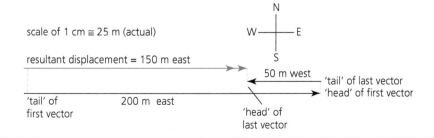

Figure 9.6

107

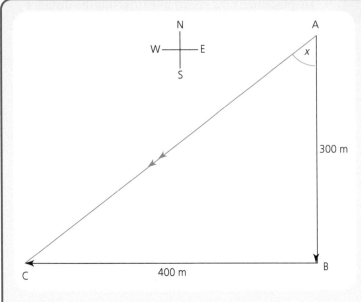

Figure 9.7

Example 2

A cyclist cycles 300 m due south, turns and then cycles 400 m due west. Calculate the cyclist's resultant displacement.

Solution

A scale vector diagram with a scale of 1 cm (diagram) ≡ 50 m (actual) is shown in Figure 9.7.

The resultant displacement (from the 'tail' of the first vector to the 'head' of the last vector) of the cyclist is represented by the line AC. The size of AC is found by measuring the length of AC with a ruler. The angle x (between the first vector and the resultant vector) is measured using a protractor.

resultant displacement = AC = 10 cm

1 cm ≡ 50 m, so

AC = 10 × 50 = 500 m

angle x = 53°

resultant displacement = 500 m at 53° W of S (or resultant displacement = 500 m on a bearing of 233)

Speed and velocity

Speed is a scalar quantity. It is the distance travelled in 1 second by an object.

Velocity is a vector quantity. It is the speed of an object in a particular direction.

average velocity = average speed in a particular direction

$$\text{average speed} = \frac{\text{distance travelled}}{\text{time taken}}$$

$$\text{average velocity} = \frac{\text{distance travelled in a particular direction}}{\text{time taken}}$$

$$\text{average velocity} = \frac{\text{displacement}}{\text{time taken}}$$

Worked examples

Example 1

A cyclist cycles 500 m due north, turns and then cycles 1200 m due east. The cyclist takes 300 s to complete this journey. For the cyclist, calculate:
a) the distance travelled
b) the average speed
c) the resultant displacement
d) the average velocity.

Solution

a) distance travelled = 500 + 1200 = 1700 m

b) average speed = $\frac{\text{distance travelled}}{\text{time taken}} = \frac{1700}{300} = 5.7\,\text{m s}^{-1}$

c) Using a scale vector diagram (Figure 9.8):
   Scale: 1 cm ≡ 150 m

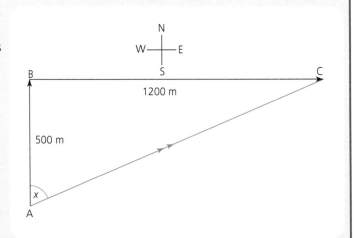

Figure 9.8

resultant displacement = AC = 8.7 cm

1 cm ≡ 150 m, so

AC = 8.7 × 150 = 1305 m

angle x = 67° (using a protractor)

resultant displacement = 1305 m at 67° E of N

d) average velocity = $\dfrac{\text{displacement}}{\text{time taken}}$

average velocity = $\dfrac{1305\,\text{m}}{300\,\text{s}}$ at 67° E of N

average velocity = 4.4 m s^{-1} at 67° E of N

Example 2

A yacht is travelling north at a speed of 2.0 m s^{-1}. The tide is moving at a speed of 1.0 m s^{-1} from west to east. Find the resultant velocity of the yacht.

Solution

The two vectors are drawn as shown in Figure 9.9.

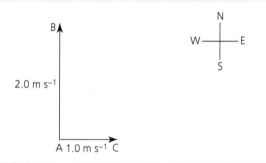

Figure 9.9

The yacht while travelling north will be pushed to the east as shown in Figure 9.10.

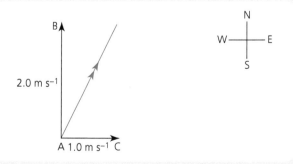

Figure 9.10

To obtain the size and direction of the resultant velocity, lines BD, CD and AD are drawn as shown in Figure 9.11.

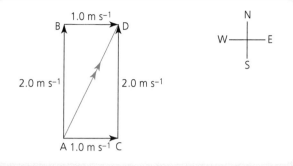

Figure 9.11

Line BD is parallel to, and has the same size as, AC, i.e. BD represents the same velocity of 1.0 m s^{-1} east as it has the same size and direction as the velocity represented by AC.

Line CD is parallel to, and has the same size as, AB, i.e. CD represents the same velocity of 2.0 m s^{-1} north as it has the same size and direction as the velocity represented by AB.

Line AD is the resultant velocity, i.e. the result of these two velocities acting on the yacht (from the 'tail' of the first vector to the 'head' of the last vector).

The resultant velocity, represented by the line AD, can be found by using either a vector diagram or a scale vector diagram.

Using a vector diagram (Figure 9.12):

$$AD^2 = AC^2 + CD^2 = 1^2 + 2^2 = 1 + 4 = 5$$

$$AD = \sqrt{5} = 2.2\,\text{m s}^{-1}$$

$$\tan x = \frac{CD}{AC} = \frac{2}{1} = 2$$

$$x = 63°$$

resultant velocity = 2.2 m s^{-1} at 63° N of E

Note that ACD and ABD are right-angled triangles so mathematics (Pythagoras and trigonometry) can be used to solve the problem.

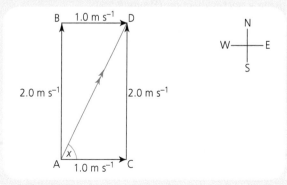

Figure 9.12

Using a scale vector diagram (Figure 9.13):

Scale: 1 cm ≡ 0.5 m s^{-1}

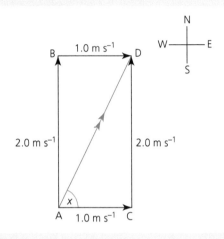

Figure 9.13

AD = 4.5 cm

1 cm ≡ 0.5 m s^{-1}, so

AD = 4.5 cm ≡ 4.5 × 0.5 = 2.25 m s^{-1} = 2.3 m s^{-1}

x = 64° (using a protractor)

resultant velocity = 2.3 m s^{-1} at 64° N of E

Note: x is the angle between the 'first' vector (in this case 1.0 m s^{-1}) and the resultant vector.

Example 3

A cruise liner is travelling at speed of 6.0 m s^{-1} due east. A girl walks at 4.0 m s^{-1} due south across the deck of the liner. Calculate the resultant velocity of the girl.

Solution

The resultant velocity, represented by the line AD as shown in Figures 9.14 and 9.15, can be found by using either a vector diagram or a scale vector diagram.

Using a vector diagram (Figure 9.14):

AD^2 = AB^2 + BD^2 = 6^2 + 4^2 = 36 + 16 = 52

AD = √52 = 7.2 m s^{-1}

$\tan x = \dfrac{BD}{AB} = \dfrac{4}{6} = 0.667$

x = 33.7° = 34°

resultant velocity = 7.2 m s^{-1} at 34° S of E

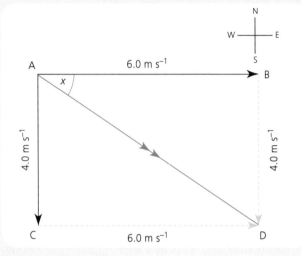

Figure 9.14

Using a scale vector diagram (Figure 9.15):

Scale: 1 cm ≡ 1 m s^{-1}

AD = 7.2 cm

1 cm ≡ 1 m s^{-1}, so

AD = 7.2 cm ≡ 7.2 × 1 = 7.2 m s^{-1}

x = 34° (using a protractor)

resultant velocity = 7.2 m s^{-1} at 34° S of E

Note: x is the angle between the 'first' vector (in this case 6.0 m s^{-1}) and the resultant vector.

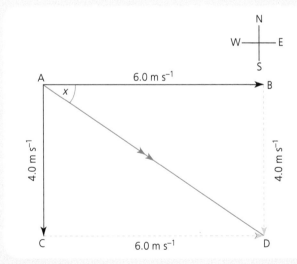

Figure 9.15

Acceleration

When the speed of an object changes so does the velocity of the object.

When the velocity of an object changes, the object is said to accelerate or to have an **acceleration**. Acceleration gives some idea as to how quickly the velocity of the object is changing.

acceleration = change in velocity in 1 second

$$\text{acceleration} = \frac{\text{change in velocity}}{\text{time taken for the velocity to change}}$$

$$\text{acceleration} = \frac{\text{final velocity} - \text{initial velocity}}{\text{time taken for the velocity to change}}$$

Provided the object is moving in a straight line, then:

$$a = \frac{v - u}{t}$$

where a = acceleration of the object, measured in metres per second squared ($m\,s^{-2}$),
v = final velocity of object, measured in metres per second ($m\,s^{-1}$),
u = initial velocity of object, measured in metres per second ($m\,s^{-1}$),
$v - u$ = change in velocity, Δv, measured in $m\,s^{-1}$,
t = time taken for the velocity to change, measured in seconds (s).

When the acceleration of an object does not change, the object is said to have a constant or uniform acceleration.

An acceleration of $5.0\,m\,s^{-2}$ means that the velocity of the object is increasing by $5.0\,m\,s^{-1}$ every second.

If the object starts from rest ($u = 0$) then:

- after 1.0 s the velocity of the object will be $5.0\,m\,s^{-1}$
- after 2.0 s the velocity of the object will be $10\,m\,s^{-1}$
- after 3.0 s the velocity of the object will be $15\,m\,s^{-1}$
- after 10 s the velocity of the object will be $50\,m\,s^{-1}$.

If the object has an initial velocity of $3.0\,m\,s^{-1}$ ($u = 3.0\,m\,s^{-1}$) then:

- after 1.0 s the velocity of the object will be $8.0\,m\,s^{-1}$
- after 2.0 s the velocity of the object will be $13\,m\,s^{-1}$
- after 3.0 s the velocity of the object will be $18\,m\,s^{-1}$
- after 10 s the velocity of the object will be $53\,m\,s^{-1}$.

When an object slows down the final velocity will be less than the initial velocity and so the change in velocity will be negative. This means that the object has a **negative acceleration** (sometimes called a deceleration).

An acceleration of $-2\,m\,s^{-2}$ means that the velocity of the object is decreasing by $2\,m\,s^{-1}$ every second.

Acceleration is a vector quantity – it has a size and a direction.

Note:

- For vector quantities: change in vector quantity = final vector − initial vector.
- For scalar quantities: change in scalar quantity = large scalar − small scalar.
- A satellite in orbit above the Earth moves at constant speed around the Earth. Although the speed of the satellite is constant, the direction of travel is constantly changing. This means the velocity of the satellite is changing and so the satellite is accelerating.

Worked examples

Example 1

A car starts from rest and travels in a straight line. The car accelerates uniformly for a time of 12 s. The speed of the car is now $25\,m\,s^{-1}$. Calculate the acceleration of the car.

Solution

$$a = \frac{v - u}{t} = \frac{25 - 0}{12} = \frac{25}{12} = 2.1\,m\,s^{-2}$$

Example 2

A lorry is travelling in a straight line. The lorry slows down uniformly from $20\,m\,s^{-1}$ to $5.0\,m\,s^{-1}$. The time taken for this is 10 s. Calculate the acceleration of the lorry.

Solution

Note: final velocity $v = 5\,m\,s^{-1}$; initial velocity $u = 20\,m\,s^{-1}$

$$a = \frac{v - u}{t} = \frac{5 - 20}{10} = \frac{-15}{10} = -1.5\,m\,s^{-2}$$

Example 3

A motorcyclist is travelling in a straight line at a speed of $16\,\mathrm{m\,s^{-1}}$. The motorcyclist now accelerates uniformly at $1.5\,\mathrm{m\,s^{-2}}$ for $4.0\,\mathrm{s}$. Calculate the final velocity of the motorcyclist.

Solution

$$a = \frac{v - u}{t}$$

$$1.5 = \frac{v - 16}{4}$$

$$v - 16 = 1.5 \times 4$$

$$v = 6 + 16 = 22\,\mathrm{m\,s^{-1}}$$

Example 4

A train is travelling in a straight line at a constant speed. The train now accelerates uniformly at $0.4\,\mathrm{m\,s^{-2}}$ for $15\,\mathrm{s}$. The train reaches a velocity of $30\,\mathrm{m\,s^{-1}}$. Calculate the initial velocity of the train.

Solution

$$a = \frac{v - u}{t}$$

$$0.4 = \frac{30 - u}{15}$$

$$30 - u = 15 \times 0.4$$

$$30 - u = 6$$

$$u = 30 - 6 = 24\,\mathrm{m\,s^{-1}}$$

Example 5

When taking-off from a runway an aeroplane travels in a straight line. The plane accelerates uniformly from rest with an acceleration of $1.8\,\mathrm{m\,s^{-2}}$. The plane reaches a speed of $50\,\mathrm{m\,s^{-1}}$. Find the time taken to reach this speed.

Solution

$$a = \frac{v - u}{t}$$

$$1.8 = \frac{50 - 0}{t}$$

$$t = \frac{50}{1.8} = 27.8\,\mathrm{s}$$

Measuring acceleration

To measure the acceleration of a vehicle travelling in a straight line down a slope requires the measurement of the initial and final speeds of the object and the time taken for the speed to change.

A card of known length is attached to the vehicle. The vehicle is placed at the top of a sloping track as shown in Figure 9.16.

Figure 9.16 Measuring acceleration using two light gates and electronic timers

The vehicle is released and moves down the slope. The card passes through a light gate connected to an electronic timer. The initial speed of the vehicle is calculated using:

$$\text{initial speed} = \frac{\text{length of card}}{\text{time on first timer}}$$

As the vehicle continues down the slope the card passes through a second light gate connected to a second timer. The final speed of the vehicle is calculated using:

$$\text{final speed} = \frac{\text{length of card}}{\text{time on second timer}}$$

The time taken for the speed to change can be measured using a stopwatch – the stopwatch is started when the card passes the first light gate and stopped when the card passes the second light gate.

The acceleration of the vehicle can then be calculated using:

$$\text{acceleration} = \frac{\text{final speed} - \text{initial speed}}{\text{time taken for the speed to change}}$$

An electronic timer/computer is available which can calculate the initial and final speeds provided it is 'programmed' with the length of the card. This type of timer/computer can also calculate the time taken for the speed to change and hence the acceleration of the vehicle.

Velocity–time graphs

In physics, a graph showing how the velocity of an object varies with time is called a velocity–time graph.

- The slope or gradient of a velocity–time graph gives the acceleration of the object.
- The area under a velocity–time graph is the displacement of the object.
- When an object is travelling in one direction in a straight line the displacement of the object is also the distance travelled by the object.

- When more than one type of motion is involved, the average velocity of the object is calculated using:

$$\text{average velocity} = \frac{\text{displacement}}{\text{time taken}}$$

- When an object moves with a constant acceleration (positive, negative or zero), the average velocity can be calculated using:

$$\frac{\text{average}}{\text{velocity}} = \frac{\text{initial velocity} + \text{final velocity}}{2} = \frac{u + v}{2}$$

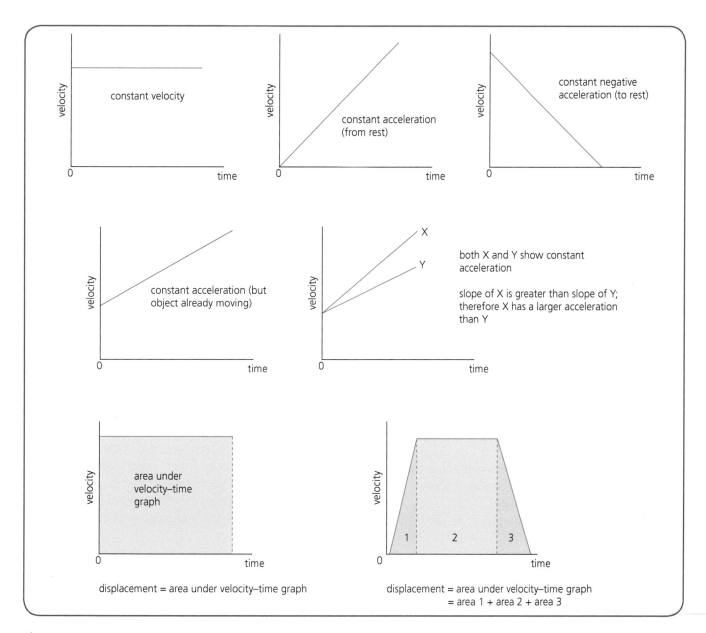

Figure 9.17

Worked examples

Example 1

An object is moving in a straight line. The graph shows how the velocity of the object varies with time.

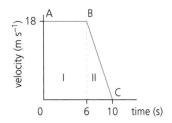

Figure 9.18

a) Describe the motion of the object between:
   i) A and B
   ii) B and C.
b) Calculate the acceleration of the object between B and C.
c) Calculate the average velocity of the object.

Solution

a) i) Constant velocity of $18\,\text{m s}^{-1}$ for $6.0\,\text{s}$
   ii) Constant negative acceleration (constant deceleration) from $18\,\text{m s}^{-1}$ to rest in $4.0\,\text{s}$

b) $a = \dfrac{v - u}{t} = \dfrac{0 - 18}{4} = \dfrac{-18}{4} = -4.5\,\text{m s}^{-2}$

   Note: $v - u = 0 - 18 = -18\,\text{m s}^{-1}$ and time for change in velocity is $4\,\text{s}$ $(10 - 6)$

c) As the object travels in a straight line and does not change direction then displacement and distance are the same.

   displacement = area under velocity–time graph
   displacement = area of I + area of II

   $$= (6 \times 18) + \left(\frac{1}{2} \times 4 \times 18\right)$$

   $$= 108 + 36 = 144\,\text{m}$$

   average velocity $= \dfrac{\text{displacement}}{\text{time taken}} = \dfrac{144}{10} = 14.4\,\text{m s}^{-1}$

Example 2

A ball rolls in a straight line down a slope. While rolling down the slope, the ball accelerates uniformly from $1.0\,\text{m s}^{-1}$ to $4.0\,\text{m s}^{-1}$ in $10\,\text{s}$. Calculate the average velocity of the ball during this time.

Solution

average velocity $= \dfrac{u + v}{2} = \dfrac{1 + 4}{2} = 2.5\,\text{m s}^{-1}$

Note: this equation may be used because acceleration is constant.

Key facts and physics equations: kinematics

- Average speed $= \dfrac{\text{distance travelled}}{\text{time taken}}$
- Average speed is measured in metres per second (m s^{-1}), distance travelled in metres (m) and time taken in seconds (s).
- A scalar quantity has size; a vector has size and direction.
- Vector quantities are: displacement, velocity, acceleration, force and weight.
- Vectors are only equal if they have the same size and the same direction.
- Average velocity $= \dfrac{\text{displacement}}{\text{time taken}}$
- Acceleration is the change in velocity of an object in 1 second.
- Acceleration $= \dfrac{\text{change in velocity}}{\text{time for change}}$

- Acceleration $= \dfrac{\text{final velocity} - \text{initial velocity}}{\text{time for change}}$

  i.e. $a = \dfrac{v - u}{t}$

- Acceleration is measured in metres per second squared (m s^{-2}), change in velocity in metres per second (m s^{-1}) and time in seconds (s).
- An acceleration of $10\,\text{m s}^{-2}$ means that the velocity of the object increases by $10\,\text{m s}^{-1}$ every second.
- An acceleration of $-5\,\text{m s}^{-2}$ means that the velocity of the object decreases by $5\,\text{m s}^{-1}$ every second.
- The slope of a velocity–time graph gives the acceleration.
- The area under a velocity–time graph gives the displacement.
- Average velocity $= \dfrac{\text{initial velocity} + \text{final velocity}}{2}$

  $= \dfrac{u + v}{2}$

  provided the acceleration is constant.

End-of-chapter questions

Information, if required, for use in the following questions can be found on the *Data Sheet* on page 170.

1 A ball is rolled across a floor. It comes to rest after 3.5 s. The ball covers a distance of 4.2 m in this time. Calculate the average speed of the ball.

2 A bus takes 30 minutes to make a journey. During the journey, the average speed of the bus is 60 km h^{-1}. Calculate the distance, in km, travelled by the bus.

3 A girl runs a 400 m race at an average speed of 6.0 m s^{-1}. Calculate the time taken for the girl to run the race.

4 State what is meant by the term *velocity*.

5 During a treasure hunt, a boy walks 300 m due north, then turns and walks 400 m due east. It takes the boy 7 minutes to do this.
   a) Calculate the distance travelled by the boy.
   b) Calculate the displacement of the boy.
   c) Calculate the average velocity of the boy.

6 A river flows due north at a constant speed of 2.5 m s^{-1}. A man rows a boat due east across the river. The speed of the boat is 3.4 m s^{-1}. Calculate the resultant velocity of the boat.

7 State what is meant by an acceleration of 5.0 m s^{-2}.

8 A skateboarder, starting from rest, accelerates uniformly down a straight concrete path. After 8.0 s, the skateboarder reaches a speed of 4.0 m s^{-1}. Calculate:
   a) the acceleration of the skateboarder
   b) the time, from starting, taken for the skateboarder to reach a speed of 6.0 m s^{-1}.

9 A marble is rolling, in a straight line, down a sloping pavement. The marble is accelerating uniformly at 0.2 m s^{-2}. It passes a line on the pavement. The speed of the marble, 5.0 s after passing the line, is 1.5 m s^{-1}. Calculate the speed of the marble as it passes the line on the pavement.

10 A car is travelling in a straight line at a speed of 30 m s^{-1}. The car now brakes and slows down uniformly at −1.5 m s^{-2}. Calculate the speed of the car 18 s after the brakes are applied.

11 To escape from the pull of the Earth, a rocket, starting from rest, has to reach a speed of 12 000 m s^{-1}. The rocket accelerates uniformly at 5.0 m s^{-2}. Calculate the time taken for the rocket to reach 12 000 m s^{-1}.

12 An object is travelling in a straight line. Figure 9.19 shows how the velocity of the object varies with time.

   a) Calculate the acceleration of the object.
   b) Calculate the average velocity of the object.

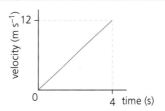

Figure 9.19

13 A vehicle is travelling in a straight line. Figure 9.20 shows how the velocity of the vehicle varies with time.

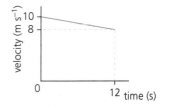

Figure 9.20

   a) Describe the motion of the vehicle.
   b) Calculate the acceleration of the vehicle.
   c) Calculate the average velocity of the vehicle for this journey.

14 A vehicle is travelling in a straight line. Figure 9.21 shows how the velocity of the vehicle varies with time.

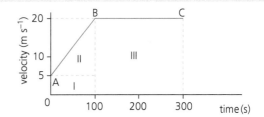

Figure 9.21

   a) Describe the motion between:
      i) the points A and B
      ii) the points B and C.
   b) Calculate the acceleration between points A and B.
   c) Calculate the displacement of the vehicle after 300 s.
   d) Calculate the average velocity of the vehicle for this journey.

10 Forces and their effects

At the end of this chapter you should be able to:

1 Describe the effects of forces in terms of their ability to change the shape, speed and direction of travel of an object.
2 Describe the use of a Newton balance to measure force.
3 State that force is a vector quantity.
4 State that forces that are equal in size but act in opposite directions on an object are called 'balanced forces' and are equivalent to no force at all.
5 State what is meant by the resultant of a number of forces.
6 Use a scale diagram, or otherwise, to find the magnitude and direction of the resultant of two forces acting at right angles to each other.
7 State that the force of friction can oppose the motion of an object.
8 Describe and explain situations in which attempts are made to increase or decrease the force of friction.
9 Distinguish between mass and weight.
10 State that weight is a force and is the pull of a planet on an object.
11 State that the gravitational field strength on a planet is the force exerted by the planet on a 1.0 kg mass.

12 Carry out calculations involving the relationship between weight, mass and gravitational field strength including situations where g is not equal to $9.8 \, \text{N} \, \text{kg}^{-1}$.
13 Explain the movement of objects in terms of Newton's first law.
14 Describe the qualitative effects of change of mass or of unbalanced force on the acceleration of an object.
15 Define the newton.
16 Use a force diagram to analyse the forces acting on an object.
17 Carry out calculations using the relationship between acceleration, unbalanced force and mass.
18 Explain the equivalence of acceleration due to gravity and gravitational field strength.
19 Explain the curved path of a projectile in terms of the force of gravity.
20 Explain how projectile motion can be treated as two independent motions.
21 Solve numerical problems using the above method for an object projected horizontally.

What can forces do?

When an object is pushed or pulled, a **force** is exerted on the object.

A force can change:

● the motion of an object, i.e. speed it up or slow it down (accelerate an object)
● the direction of a moving object
● the shape of an object.

These changes depend upon how large a force is applied to the object.

Measuring force

A spring can be used to measure force:

● A spring stretches evenly when acted on by the same size of force. The stretch is directly proportional to the force, i.e. doubling the force doubles the size of the stretch.
● A spring returns to its original length when the force is removed.

A Newton (or spring) balance (Figure 10.1) uses a spring to measure force. The unit of force is the newton (N).

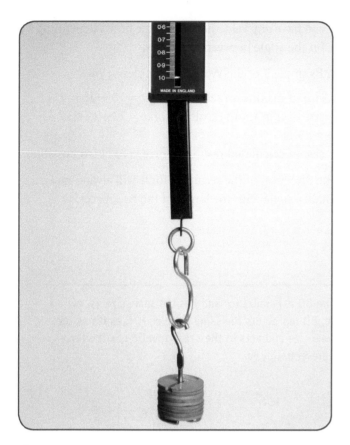

Figure 10.1 A Newton balance

Force: a vector quantity

Figure 10.2 shows forces acting on objects X and Y.

Figure 10.2 Forces acting on objects X and Y

Since force is a vector quantity, it requires a size and a direction to describe it.

Taking forces to the right as being positive, then:

- force P = +10 N = 10 N (10 N to the right)
- force Q = +6 N = 6 N (6 N to the right)
- force R = +10 N = 10 N (10 N to the right)
- force S = −6 N (6 N to the left)

Two forces are equal only when they have the same size and act in the same direction.

- Forces P and R are equal – they have the same size and act in the same direction.
- Forces Q and S are not equal – they have the same size but they act in opposite directions.

When two or more forces act on an object, the combined effect of these forces depends on their size and their direction. A single force called the **resultant force** or **unbalanced force** can replace them.

The unbalanced (resultant) force is found by 'adding up' all the forces acting on an object. The direction of each force must be taken into account.

Balanced forces

Figure 10.3 shows two forces acting on an object R.

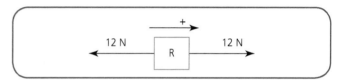

Figure 10.3 Forces acting on object R

The two forces are equal in size but act in opposite directions. The forces cancel each other out. This is equivalent to no force acting on the object.

(The single force or resultant force or unbalanced force acting on R = (+12) + (−12) = 0 N.)

The forces are said to be balanced.

Balanced forces:

- have the same size
- act in opposite directions
- act on the same object.

Unbalanced forces

Figure 10.4 shows two forces acting on objects P and Q.

Figure 10.4 Forces acting on objects P and Q

For object P, the two forces are not equal in size but do act in the same direction:

$$\text{unbalanced force acting on P} = (+6) + (+4)$$
$$= (+10)\,\text{N}$$
$$= 10\text{N} \ (10\text{N to the right})$$

117

For object Q, the two forces are not equal in size and act in opposite directions:

$$\text{unbalanced force acting on Q} = (+6) + (-4)$$
$$= (+2)\,\text{N}$$
$$= 2\,\text{N} \ (2\,\text{N to the right})$$

The size of the resultant of two forces:

- is a maximum when the forces act in the same direction (are added together)
- is a minimum when the forces act in opposite directions (are subtracted)

- can have any value between these limits depending on the angle between the forces.

In Example 2 of the Worked examples below:

- size of maximum resultant force = 70 + 25 = 95 N
- size of minimum resultant force = 70 + (−25) = 45 N
- size of calculated resultant force = 74 N (so possible)

The direction of the resultant force will always make a smaller angle with the larger of the two forces.

Worked examples

Example 1

Two forces act on an object as shown in Figure 10.5. What is the unbalanced force acting on the object?

Figure 10.5

Solution

Taking vectors to the right as positive:
    unbalanced force = (+18) + (−6) = (+12) = 12 N

This means that the unbalanced force acting on the object has a size of 12 N to the right.

Example 2

Two forces act on an object as shown in Figure 10.6. Calculate the resultant force acting on the object.

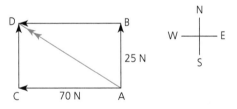

Figure 10.6

Solution

While being pulled to the left the object will also be pulled upwards, as shown in Figure 10.7.

To obtain the size and direction of the resultant force, lines BD, CD and AD are drawn as shown in Figure 10.8.

Line BD is parallel to, and has the same size as, AC, i.e. BD represents the same force of 70 N as it has the same size and acts in the same direction as the force represented by AC.

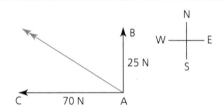

Figure 10.7

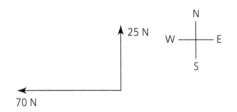

Figure 10.8

Line CD is parallel to, and has the same size as, AB, i.e. CD represents the same force of 25 N as it has the same size and acts in the same direction as the force represented by AB.

Line AD is the resultant force, i.e. the result of these two forces acting on the object.

The resultant force is found by drawing a line from A to D (from the 'tail' of the first vector to the 'head' of the last vector). The resultant force, represented by the line AD, can be found using either a vector diagram (Figure 10.9) or a scale vector diagram (Figure 10.10).

Using a vector diagram:

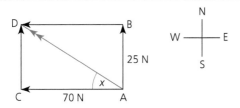

Figure 10.9 Vector diagram

$AD^2 = AC^2 + CD^2 = 70^2 + 25^2 = 4900 + 625$

$AD^2 = 5525$

$AD = \sqrt{5525} = 74\,N$

$\tan x = \dfrac{CD}{AC} = \dfrac{25}{70} = 0.357$

angle $x = 19.7°$

resultant force = 74 N at 20° N of W

Using a scale vector diagram:

Scale: 1 cm ≡ 10 N

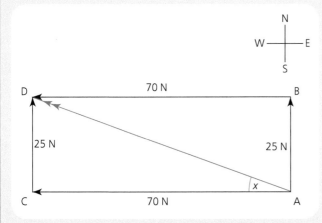

Figure 10.10 Scale vector diagram

AD = 7.3 cm

1 cm ≡ 10 N, so

AD = 7.3 × 10 = 73 N

angle $x = 20°$ (using a protractor)

resultant force = 73 N at 20° N of W

Note: angle x is the angle between the 'first' vector (in this case 70 N) and the resultant vector.

Example 3

A canal barge is pulled by two forces as shown in Figure 10.11. Calculate the resultant force produced by these two forces.

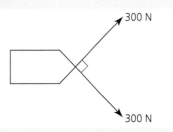

Figure 10.11

Solution

The resultant force can be found using either a vector diagram or a scale vector diagram.

Using a vector diagram:

A diagram of forces (not to scale) is shown in Figure 10.12.

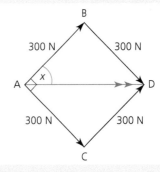

Figure 10.12 Vector diagram

$AD^2 = AC^2 + CD^2 = 300^2 + 300^2$

$AD^2 = 90\,000 + 90\,000 = 180\,000$

$AD = \sqrt{180\,000} = 424\,N$

$\tan x = \dfrac{CD}{AC} = \dfrac{300}{300} = 1.0$

angle $x = 45°$

resultant force = 424 N at 45° from either force

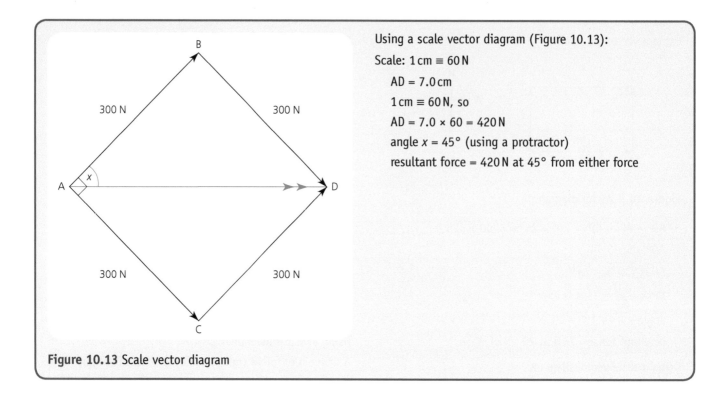

Using a scale vector diagram (Figure 10.13):

Scale: 1 cm ≡ 60 N

AD = 7.0 cm

1 cm ≡ 60 N, so

AD = 7.0 × 60 = 420 N

angle x = 45° (using a protractor)

resultant force = 420 N at 45° from either force

Figure 10.13 Scale vector diagram

Frictional forces

An object is pushed along a bench (Figure 10.14). When the pushing force is removed, the object slows down and stops. This can only happen if there is a force acting on the object. This force is due to the particles of the object and the bench sliding across one another. This causes a force called the **force of friction**. Friction is a force that usually tries to oppose the motion of an object – tries to stop the object from moving. When an object is moving to the right, the force of friction acts towards the left.

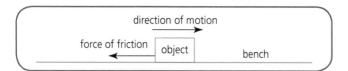

Figure 10.14 An object moving along a bench

Increasing frictional forces

When a moving object has to be slowed down, the frictional force on the object must be increased. In a bicycle brake this is done by squeezing rubber blocks against the metal wheel rim, thus increasing the frictional force. In a car, applying the brakes causes a pad to rub against a disc (part of the wheel) on each of the wheels, thus increasing the force of friction.

Some high-performance aircraft use a parachute when landing to slow them down rapidly (see Figure 10.15). The large area of the parachute greatly increases the frictional force caused by air resistance.

Figure 10.15 The large surface area of a parachute creates a large resistive force

Reducing frictional forces

Frictional forces can cause a great deal of energy to be wasted when an object is moving. In this case it is useful to reduce frictional forces.

For example:

- Lubrication, which involves putting a fluid (a liquid or gas) between the two surfaces that slide across each other:
  - Oil is used to lubricate the moving parts of a car engine. The oil does not evaporate at high temperatures.
  - In a linear air track, a thin layer of air is used to separate the vehicle from the track. Hovercraft (see Figure 10.16) use a layer of air to separate the vehicle from the ground.

Figure 10.16 A hovercraft

- Streamlining the shape of an object. The object offers as little resistance to the air as possible. This is particularly important in vehicles that are required to move at high speeds (see Figure 10.17).

Figure 10.17 High-speed trains need to be streamlined

Mass

Mass is the quantity of matter forming an object. It depends on the number and type of particles that make up the object. Mass is measured in kilograms (kg).

Provided the number and type of particles making up an object does not change, the mass of the object remains the same no matter where it is taken.

A hammer has a mass of 0.5 kg on the Earth. Since the hammer is made up of a certain number and type of particles, the same hammer on the Moon will still have a mass of 0.5 kg. Taking the same hammer into outer space will not change the number and type of particles, so it will still have a mass of 0.5 kg.

Force of gravity

The **force of gravity** can be defined as follows:

- force of gravity = downward pull of a planet on an object
- force of gravity = force (or pull) exerted by a planet on an object
- force of gravity = gravitational force on an object
- force of gravity = weight of an object.

Weight

The **weight** of an object depends on:

- mass – the number and type of particles that make up the object
- location (where it is) – for most situations this will be the surface of the Earth. However, a 0.5 kg hammer has a smaller weight on the surface of the Moon than it does on the Earth.

Since weight is a force, it is measured in newtons (N).

Gravitational field strength

For any object on a planet:

$$\frac{\text{weight of object}}{\text{mass of object}} = \frac{W}{m} = \text{constant}$$

The constant is called the **gravitational field strength** (symbol g).

Gravitational field strength (g) is the force exerted by a planet on a 1.0 kg mass. It is measured in newtons per kilogram ($N\,kg^{-1}$).

This is normally written as:

$$\begin{array}{ccc} \text{weight} \\ \text{of object} \end{array} = \begin{array}{c} \text{mass of} \\ \text{object} \end{array} \times \begin{array}{c} \text{gravitational field} \\ \text{strength} \end{array}$$

$$W = mg$$

where W = weight of object, measured in newtons (N), m = mass of object, measured in kilograms (kg), g = gravitational field strength, measured in newtons per kilogram ($N\,kg^{-1}$).

The gravitational field strength for the Earth is $9.8\,N\,kg^{-1}$ ($9.8\,N$ for every kg). This means that on Earth, a 1.0 kg mass has a weight of 9.8 N and a 10 kg mass has a weight of 98 N.

A table of gravitational field strengths is given on the *Data Sheet* on page 170.

Worked example

Example

The mass of a hammer is 0.5 kg. Calculate the weight of the hammer on the surface of:
a) the Earth
b) the Moon
c) Jupiter.

Solution

a) On the Earth:
   $W = mg = 0.5 \times 9.8 = 4.9\,N$
   Note: g for Earth from *Data Sheet*
b) On the Moon:
   $W = mg = 0.5 \times 1.6 = 0.8\,N$
   Note: g for Moon from *Data Sheet*
c) On Jupiter:
   $W = mg = 0.5 \times 23 = 11.5\,N$
   Note: g for Jupiter from *Data Sheet*

Newton's first law

Newton's first law states that:

'An object will remain at rest or move at constant speed in a straight line (constant velocity) unless acted on by an unbalanced force.'

This means that if an object is at rest or is moving at a constant speed in a straight line, then there is either no force acting on the object or the forces acting on the object are balanced.

Worked example

Example

A car is travelling in a straight line along a level road at a constant speed. Draw and name the forces acting horizontally on the car.

Solution

Since the car is travelling at constant speed in a straight line Newton's first law applies – either there are no forces acting on the car or the forces acting on the car are balanced. However, a car will only travel at a constant speed along a level road when the engine provides a force to balance out the resistive forces acting on the car (Figure 10.19).

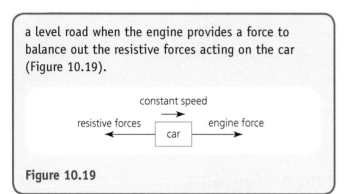

Figure 10.19

Forces and supported objects

A book is at rest on a table (Figure 10.20). The book has a mass, m.

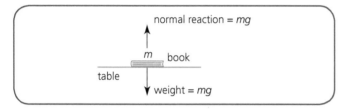

Figure 10.20 Vertical forces acting on a book

- The book's weight ($= mg$) acts downwards.
- The table exerts an upward force, called the normal reaction, on the book.
- The normal reaction is equal in size to the weight of the book, but acts in the opposite direction. The forces are balanced (one is a 'positive' force and the other is a 'negative' force).

A lift is at rest (Figure 10.21). The lift is supported by a cable. The lift has a mass, m.

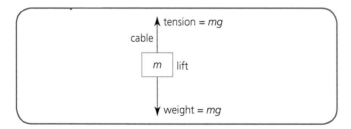

Figure 10.21 Vertical forces acting on a lift

- The lift's weight ($= mg$) acts downwards.
- The lift cable exerts an upward force on the lift called the tension.
- The tension is equal in size to the weight of the lift, but acts in the opposite direction. The forces are balanced.

Note: if the lift were moving upwards or downwards at constant speed, then from Newton's first law the forces acting on the lift would still be balanced, i.e. the tension would be equal in size but act in the opposite direction to the weight of the lift.

Newton's second law

Figure 10.22 shows a vehicle on a level linear air track. A card of known length is attached to the vehicle. As the vehicle moves along the track the card passes through a number of light gates. The light gates are connected to electronic timers. These timers calculate the acceleration of the vehicle as it moves along the track. A mass, at one end of the linear air track, is attached to the vehicle by a light string and a pulley.

Figure 10.22 Measuring acceleration of a vehicle on a linear air track

When the mass is allowed to fall it applies a constant unbalanced force to the vehicle. The acceleration of the vehicle is found to be constant and in the same direction as the unbalanced force. Further investigation shows that:

- the acceleration, a, of the vehicle doubles as the unbalanced force, F_{un}, on the vehicle doubles (a is directly proportional to F_{un}, mass of vehicle constant)
- the acceleration, a, of the vehicle halves as the mass, m, of the vehicle doubles (a is indirectly proportional to m, unbalanced force constant).

Hence, when a constant unbalanced force is applied to an object, the object will move with a constant acceleration in the direction of the unbalanced force.

$$\text{unbalanced force on object} = \text{mass of object} \times \text{acceleration of object}$$

$$F_{un} = ma$$

where F_{un} = unbalanced force on object, measured in newtons (N),
m = mass of object, measured in kilograms (kg),
a = acceleration of object, measured in metres per second squared (m s^{-2}).

1.0 N is the constant unbalanced force that gives a 1.0 kg mass a constant acceleration of 1.0 m s^{-2}.

The acceleration of an object can be changed:

- by changing the size of the unbalanced force on the object (increasing F_{un} increases a), or
- by changing the mass of the object (increasing m decreases a).

Worked examples

Example 1

The acceleration of an object is 1.5 m s^{-2}. The mass of the object is 2.5 kg. Calculate the unbalanced force acting on the object.

Solution

$$F_{un} = ma$$
$$F_{un} = 2.5 \times 1.5$$
$$F_{un} = 3.75 \text{ N}$$

Example 2

The mass of an object is 5.0 kg. The unbalanced force acting on the object is 35 N. Calculate the acceleration of the object.

Solution

$$F_{un} = ma$$
$$35 = 5 \times a$$
$$a = \frac{35}{5} = 7.0 \text{ m s}^{-2}$$

Example 3

An unbalanced force of 3.5 kN is applied to a vehicle. The acceleration of the vehicle is 0.2 m s^{-2}. Calculate the mass of the vehicle.

Solution

Note: $3.5 \text{ kN} = 3.5 \times 10^3 \text{ N} = 3500 \text{ N}$

$$F_{un} = ma$$
$$3.5 \times 10^3 = m \times 0.2$$
$$m = \frac{3.5 \times 10^3}{0.2} = 17\,500 \text{ kg}$$

Force diagrams

- When dealing with problems relating to forces, always draw a force diagram.
- Mark in all the forces acting on the object.
- Calculate the unbalanced (resultant) force acting on the object.
- Use the equation $F_{un} = ma$.

Worked examples

Example 1

The mass of an object is 4.0 kg. A force of 15 N is applied to an object. The frictional force acting on the object is 5.0 N. Calculate the acceleration of the object.

Solution

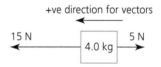

Figure 10.23

$F_{un} = 15 + (-5) = +10 = 10 \text{ N}$ (meaning 10 N to the left)

$$F_{un} = ma$$
$$10 = 4 \times a$$
$$a = \frac{10}{4} = 2.5 \text{ m s}^{-2} \text{ (meaning } 2.5 \text{ m s}^{-2} \text{ to the left)}$$

Example 2

The total mass of a lorry is 11 500 kg. The lorry is initially at rest. The engine of the lorry now applies a forward force of 5.1 kN and the lorry moves off in a

straight line. When moving, a constant frictional force of 500 N acts on the lorry.

a) Calculate the acceleration of the lorry.
b) Calculate the time taken, from moving off, for the lorry to reach a speed of 12 m s^{-1}.

Solution

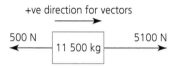

+ve direction for vectors

500 N ← | 11 500 kg | → 5100 N

Figure 10.24

a) Note: 5.1 kN = 5.1 × 10^3 N = 5100 N

unbalanced force = 5100 + (−500) = 4600 N

$$F_{un} = ma$$

$$4600 = 11\,500 \times a$$

$$a = \frac{4600}{11\,500} = 0.4\,\text{m s}^{-2} \text{ (meaning 0.4 m s}^{-2} \text{ to the right)}$$

b) $$a = \frac{v - u}{t}$$

$$0.4 = \frac{12 - 0}{t}$$

$$t = \frac{12}{0.4} = 30\,\text{s}$$

Physics beyond the classroom

Seat belts

When a moving car brakes suddenly, any unrestrained object will continue to move at the car's original speed (Newton's first law) until, most likely, it collides with some part of the interior. This will probably cause damage or injury to the object.

A seat belt applies a force in the opposite direction of motion and this rapidly decelerates the wearer. The webbing straps are designed to have a certain amount of 'give' so that the sudden force applied to the person does not cause injury.

Air bags

Air bags are designed to provide a cushion between you and the dashboard or steering wheel of a car. This increases the time taken to bring you to rest. This means that the deceleration is smaller in size and so you experience a smaller unbalanced force, which means you are less likely to be injured.

Head restraints

These are designed to reduce neck injury. They are particularly useful if your car is hit from behind. As the car is pushed forward, the back of your seat pushes you forward. Without a head restraint, your head will tend to remain where it is, while the rest of your body moves forward. This can cause a 'whiplash' injury.

Crumple zones

In a collision, the crumple zones in a car allow the car to come to rest over a longer time. The size of the deceleration of the car is reduced. This means that there is a smaller unbalanced force acting on the car and its occupants, causing less injury.

Acceleration due to gravity

Consider three objects, of different mass, that are allowed to fall vertically downwards near the surface of the Earth as shown in Figure 10.25. The effects of air resistance can be ignored.

When the effects of air resistance are ignored, the only force acting on each object is the weight of the object. This is also the unbalanced force on the object. The objects will move with a constant acceleration of 9.8 m s^{-2} downwards.

In the absence of frictional forces, all objects falling freely have the same downward acceleration no matter what the mass of the object is. This acceleration is called the **acceleration due to gravity** (g).

The acceleration due to gravity (g) on the Earth is 9.8 m s^{-2}, which is the same value as the gravitational field strength on the Earth.

On any planet, the gravitational field strength and the acceleration due to gravity have the same value:

gravitational field strength (N kg^{-1}) = g = acceleration due to gravity (m s^{-2})

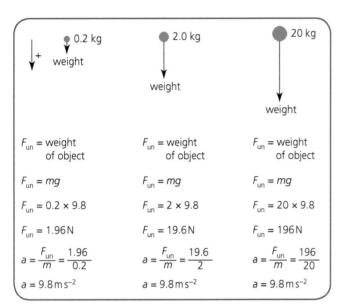

Figure 10.25

Projectile motion

Consider an object taken to three different planets, X, Y and Z. Information about these planets, and what happens to the object, is given in Figure 10.26.

The combined effect of the two motions on planet Z is that the object will follow a curved path downwards. This is called **projectile motion**.

The drawing of a multi-image photograph shown in Figure 10.27 shows two balls that were released at the same time. One ball was allowed to fall vertically downwards while the other ball was projected horizontally.

Notice that at each horizontal image in Figure 10.27, the vertical positions of both balls are the same. Both balls must therefore have the same vertical motion. An analysis of the vertical distances travelled shows that the vertical motion is **constant acceleration of $9.8\,\mathrm{m\,s^{-2}}$.**

Planet X	Planet Y	Planet Z
$g = 0\,\mathrm{N\,kg^{-1}}$	$g = 9.8\,\mathrm{N\,kg^{-1}}$	$g = 9.8\,\mathrm{N\,kg^{-1}}$
air resistance = 0 N	air resistance = 0 N	air resistance = 0 N
Object is projected horizontally with a speed of $20\,\mathrm{m\,s^{-1}}$.	Object is dropped vertically from rest.	Object is projected horizontally with a speed of $20\,\mathrm{m\,s^{-1}}$.
	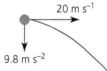	
Since there are no forces acting on it, the object will continue to move at $20\,\mathrm{m\,s^{-1}}$ horizontally (constant velocity).	The weight of the object is the only force acting on the object.	Object will keep moving at a constant speed of $20\,\mathrm{m\,s^{-1}}$ horizontally (as on planet X).
	Object will have a constant acceleration of $9.8\,\mathrm{m\,s^{-2}}$ downwards.	At the same time the object will accelerate vertically downwards with a constant acceleration of $9.8\,\mathrm{m\,s^{-2}}$ (as on planet Y).

Figure 10.26

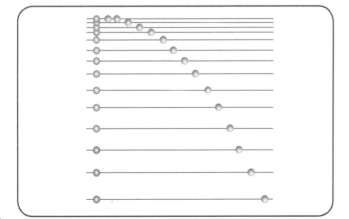

Figure 10.27 Drawing from a multi-image photograph of two balls falling to the ground

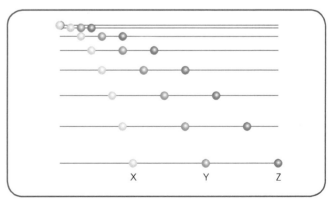

Figure 10.28 Drawing from a multi-image photograph showing three balls being projected horizontally

Figure 10.28 shows three identical balls, X, Y and Z, that have been projected horizontally at different speeds.

The vertical motion of the balls is unaffected by the horizontal motion.

All three balls have the same vertical motion, i.e. constant acceleration of 9.8 m s^{-2}.

In the same time, Z travels the greatest horizontal distance – Z has the greatest horizontal velocity.

The motion of a projectile seems complex but it can be divided into two simple motions that are independent of each other (provided the effects of air resistance are ignored):

- a horizontal motion of constant velocity, which has the same value as the initial horizontal velocity
- a vertical motion of a constant acceleration of 9.8 m s^{-2} starting from rest.

For an object projected horizontally, the velocity–time graphs for the horizontal and vertical motions are shown in Figure 10.29.

When solving problems on projectiles, the motion in the horizontal direction is treated separately from the motion in the vertical direction.

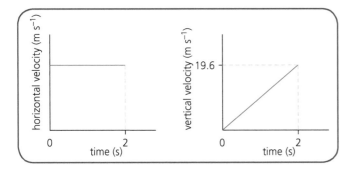

Figure 10.29

For the horizontal velocity-time graph:

- Since the object is travelling horizontally (in one direction), the horizontal displacement is also the horizontal distance travelled by the object.
- horizontal displacement = area under horizontal velocity-time graph

  horizontal displacement = range of object

For the vertical velocity-time graph:

- The slope or gradient of the graph gives the (vertical) acceleration of the object. For the Earth this is 9.8 m s^{-2}.

- Since the object is travelling vertically (in one direction), the vertical displacement is also the vertical distance travelled by the object (height).
- height = vertical displacement = area under vertical velocity–time graph

or

height = average vertical velocity × time
$$= \frac{(u + v)}{2} \times \text{time}$$

Worked examples

Example 1

A ball is thrown horizontally at 8.0 m s^{-1} from a high tower. The ball hits the ground 4.0 s later. The effects of air resistance can be ignored. Calculate:
a) the horizontal distance travelled by the ball in 4.0 s
b) the vertical speed of the ball just as it is about to hit the ground
c) the velocity of the ball just as it is about to hit the ground.

Solution

a) Note: horizontal motion is a constant speed of 8.0 m s^{-1}
$$\text{horizontal speed} = \frac{\text{horizontal distance}}{\text{time taken}}$$
$$8 = \frac{\text{horizontal distance}}{4}$$
horizontal distance = 8 × 4 = 32 m

b) Note: vertical motion is a constant acceleration of 9.8 m s^{-2} from rest
$$a = \frac{v - u}{t}$$
$$9.8 = \frac{v - 0}{4}$$
$$v = 9.8 \times 4 = 39.2 \text{ m s}^{-1}$$

c) Using a vector diagram (Figure 10.30):
$$AD^2 = AC^2 + CD^2 = 39.2^2 + 8^2 = 1537 + 64$$
$$AD^2 = 1601$$

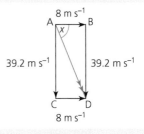

Figure 10.30 Vector diagram

$AD = \sqrt{1601} = 40\,\text{m}\,\text{s}^{-1}$

$\tan x = \dfrac{BD}{AB} = \dfrac{39.2}{8} = 4.9$

$x = 78.5°$

resultant velocity = $40\,\text{m}\,\text{s}^{-1}$ at 79° from the horizontal

Using a scale vector diagram (Figure 10.31):

Scale: $1\,\text{cm} \equiv 4\,\text{m}\,\text{s}^{-1}$

$AD = 9.9\,\text{cm}$

$1\,\text{cm} \equiv 4\,\text{m}\,\text{s}^{-1}$, so

$AD = 9.9 \times 4 = 39.6\,\text{m}\,\text{s}^{-1}$

$x = 79°$ (using a protractor)

resultant velocity = $40\,\text{m}\,\text{s}^{-1}$ at 79° from the horizontal

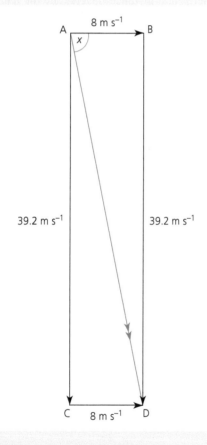

Figure 10.31 Scale vector diagram

Example 2

A ball is projected horizontally from the window of a house. The ball hits the ground 2.0 s later at a horizontal distance of 15 m from the foot of the house. The effects of air resistance can be ignored. Calculate:
a) the horizontal speed of projection of the ball
b) the vertical speed of the ball just before it hits the ground
c) the height from which the ball is projected.

Solution

a) horizontal speed = $\dfrac{\text{horizontal distance}}{\text{time taken}}$

$= \dfrac{15}{2} = 7.5\,\text{m}\,\text{s}^{-1}$

b) $a = \dfrac{v - u}{t}$

$9.8 = \dfrac{v - 0}{2}$

$v = 19.6\,\text{m}\,\text{s}^{-1}$

c) height = vertical distance travelled by ball

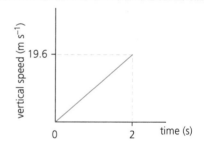

Figure 10.32

height = area under vertical speed–time graph

height = area under graph = $\dfrac{1}{2} \times 2 \times 19.6 = 19.6\,\text{m}$

Alternatively:

average vertical speed = $\dfrac{u + v}{2}$

Note: this equation may be used because acceleration is constant.

average vertical speed = $\dfrac{0 + 19.6}{2} = 9.8\,\text{m}\,\text{s}^{-1}$

height = vertical distance travelled

height = average vertical speed × time taken

height = 9.8×2

height = $19.6\,\text{m}$

Newton's satellite

Newton considered the motion of the Moon around the Earth. His 'thoughts' went like this:

'Suppose a bullet is fired horizontally from a gun situated on top of a high mountain. The bullet will have two motions that occur at the same time (if the effects of air resistance can be ignored). These are:

- a horizontal motion with a constant velocity
- a vertical motion with a uniform acceleration of *g* due to the gravitational pull of the Earth on the bullet.

The bullet will follow a curved path (a projectile) and will hit the ground no matter what the speed of projection is, as long as the Earth is flat. However the approximation of a 'flat' Earth is only valid over a limited range. For a 'round' Earth, when the horizontal distance travelled by the bullet is large, it becomes important to take the curvature of the Earth into account (Figure 10.33).

The bullet will only follow the path AB if there is no force of gravity present. Since there would be no forces acting on the bullet, it would travel at constant speed in a straight line (Newton's first law).

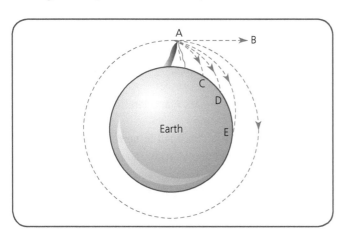

Figure 10.33 Motion of a bullet projected from a high mountain

However, due to the force of gravity, the bullet falls continuously below the line AB. When the bullet falls below the line AB faster than the surface of the Earth curves away from under it, the bullet will still hit the surface of the Earth at, for example, points C, D or E, depending on the speed of the bullet.

When the bullet is fired at just the right speed so that it falls vertically as quickly as the surface of the Earth curves away under it, then the bullet would always be at the same height above the surface of the Earth. The bullet would never reach the surface of the Earth, but would circle the Earth at a constant height – a satellite, just like the Moon.'

Resistive forces in a fluid

A fluid is a liquid or a gas. When an object moves through a fluid the resistive force of the fluid on the object is not constant but depends on:

- the speed of the object – as the speed of the object in the fluid increases, the resistive force on the object increases
- the 'thickness' or viscosity of the fluid – water is more viscous than air, so it is harder to move through water than air (as water provides a greater resistive force on an object than air).

The motion of an object falling through a fluid can be divided into three parts (see Figure 10.34).

1 Initially, the only force acting on the object is its weight – the unbalanced force. The object falls with a constant acceleration of $9.8\,\mathrm{m\,s^{-2}}$ ($F_{un} = W$).
2 After a very short time the resistive force (F_r) of the fluid begins to act. This force opposes the motion of the object and it increases as the speed of the object increases. There is a smaller unbalanced force $F_{un} = W + (-F_r)$. The acceleration of the object decreases.
3 Finally the resistive force has increased to the same size as the weight of the object – balanced forces. The object now falls at a constant speed in a straight line having reached its terminal velocity.

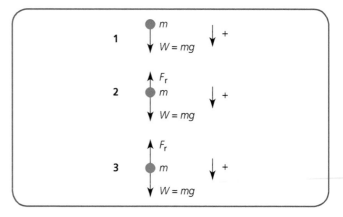

Figure 10.34 Motion of an object in a fluid

Free-falling parachutist

The graph in Figure 10.35 shows how the vertical velocity of a parachutist varies with time from the instant of 'stepping out' of the plane until hitting the ground.

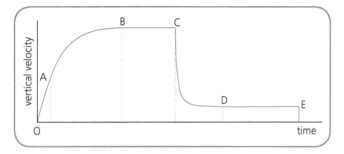

Figure 10.35 Vertical velocity against time for a parachutist (not to scale)

- From O to A, the parachutist initially has an acceleration of $9.8\,\mathrm{m\,s^{-2}}$ as the only vertical force acting is the weight of the parachutist.

- From A to B, the resistive force of the air acts on the parachutist. The resistive force increases as the speed increases. There is still an unbalanced force and an acceleration but both decrease as the parachutist's speed increases [$F_{un} = W + (-F_r)$].
- From B to C, the parachutist has reached their terminal velocity. The forces acting on the parachutist are balanced (the weight and the resistive force are equal in size but opposite in direction).
- At C, the parachute is opened. The resistive force of the air, due to the larger area of the parachute, is initially much larger than the weight.
- From C to D, the parachutist's speed decreases. The resistive force also decreases as the speed decreases.
- From D to E, the weight and the resistive force are again balanced. The speed of the parachutist is much lower than at C due to the larger area of the parachute.
- At E the parachutist hits the ground.

Key facts and physics equations: forces and their effects

- A force can change the speed, direction of travel and shape of an object.
- A Newton balance is used to measure force in newtons (N).
- Force is a vector quantity.
- Forces that are equal in size but act in opposite directions on an object are called balanced forces and are equivalent to no force acting on the object.
- Mass is the quantity of matter making up an object; mass remains the same.
- Weight is a force and is a planet's pull on an object.
- Weight = mass × gravitational field strength, i.e. $W = mg$.
- Weight is measured in newtons (N), mass in kilograms (kg) and gravitational field strength in newtons per kilogram ($\mathrm{N\,kg^{-1}}$).

- Gravitational field strength on a planet is the weight of a 1.0 kg mass and is constant for a planet.
- Newton's first law: an object will remain at rest or move at constant speed in a straight line unless acted on by an unbalanced force.
- Newton's second law:
  unbalanced force = mass × acceleration
  i.e. $F_{un} = ma$
- Unbalanced force is measured in newtons (N), mass in kilograms (kg) and acceleration in metres per second squared ($\mathrm{m\,s^{-2}}$).
- Acceleration due to gravity and gravitational field strength for a planet have the same value.
- Acceleration due to gravity on Earth = $9.8\,\mathrm{m\,s^{-2}}$ = gravitational field strength = $9.8\,\mathrm{N\,kg^{-1}}$.
- For an object projected horizontally above the surface of the Earth, the horizontal and vertical motions are independent of each other; horizontal motion is constant speed, vertical motion is constant acceleration of $9.8\,\mathrm{m\,s^{-2}}$ from rest – if the effects of air resistance are ignored.

End-of-chapter questions

Information, if required, for use in the following questions can be found on the *Data Sheet* on page 170.

1 Two forces act on a crate as shown in Figure 10.36. Calculate the resultant force produced by these two forces.

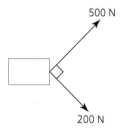

500 N

200 N

Figure 10.36

2 A car is travelling in a straight line along a road. The engine force and the resistive force are the only forces acting horizontally on the car.
   a) Draw a labelled diagram showing the forces acting horizontally on the car.
   b) How do the forces compare:
      i) when the car is travelling at a constant speed
      ii) when the car has a constant positive acceleration.

3 A space probe is sent from the Earth and lands on Mars. On Earth the mass of the probe is 170 kg.
   a) What is the mass of the probe on Mars?
   b) Calculate the weight of the probe on the surface of
      i) the Earth
      ii) Mars.

4 Figure 10.37 shows the four forces acting on a flying aircraft.

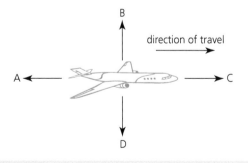

B

direction of travel

A

C

D

Figure 10.37

a) Match each of the letters A, B, C and D with the correct force from the following list:
   air resistance; engine force; lift force (from wings); weight.
b) The aircraft is flying at a constant speed in a straight line and at a constant height above the ground. State how the following pairs of forces compare in size and direction:
   i) A and C
   ii) B and D.

5 State what is meant by *gravitational field strength*.

6 A lampshade has a mass of 1.5 kg. It is suspended from a ceiling by a flex.
   a) Calculate the weight of the lampshade.
   b) What is the size and direction of the tension in the flex?

7 The mass of a bus is 5000 kg. The bus has an acceleration of 0.12 m s⁻². Calculate the unbalanced force acting on the bus.

8 An unbalanced force of 0.2 N acts on a linear air track vehicle. The mass of the vehicle is 400 g. Calculate the acceleration of the vehicle.

9 An unbalanced force of 5.4 kN acts on a light plane. The acceleration of the plane is 2.7 m s⁻². Calculate the mass of the plane.

10 The mass of a trolley is 0.3 kg. A force of 2.0 N is applied to the trolley. The acceleration of the trolley is 4.0 m s⁻².
   a) Calculate the unbalanced force acting on the trolley.
   b) Calculate the size of the resistive force acting on the trolley.

11 The mass of a helicopter is 1.3×10^4 kg. The rotor blades exert an upward force of 1.6×10^5 N on the helicopter just as it lifts off.
   a) Calculate the weight of the helicopter.
   b) Draw a box to represent the helicopter. Show and name all of the forces acting vertically on the helicopter just as it lifts off.
   c) Calculate the acceleration of the helicopter just as it lifts off.

12 The mass of a car is 1100 kg. The engine of the car applies a forward force of 500 N. At a certain time the total resistive force on the car is 150 N.

    a) Calculate the acceleration of the car at this time.

    b) Later, the car is moving at a constant speed in a straight line. The car's engine still applies a forward force of 500 N. Explain in terms of forces, why the car is travelling at a constant speed.

13 A woman is holding a glass. The glass has a mass of 200 g.

    a) Calculate the weight of the glass.

    b) The woman accidentally drops the glass. The glass falls to the floor. What is the acceleration of the glass?

14 A ball is projected horizontally from a high tower with a speed of $18\,\mathrm{m\,s^{-1}}$. The ball hits the ground 3.0 s later. The effects of air resistance can be ignored.

    a) Calculate the horizontal distance travelled by the ball in 3.0 s.

    b) Calculate the vertical speed of the ball just as it is about to hit the ground.

    c) Calculate the vertical height the ball falls through.

Newton's third law and energy

At the end of this chapter you should be able to:
1 State Newton's third law.
2 Identify 'Newton pairs' in situations involving several forces.
3 State that work done is a measure of the energy transferred.
4 Carry out calculations involving the relationship between work done, force applied and distance moved by the force.
5 Carry out calculations involving the relationship between change in gravitational potential energy, mass, gravitational field strength and change in height.
6 Carry out calculations involving the relationship between kinetic energy, mass and velocity.
7 Carry out calculations involving the relationship between work done, power and time.
8 Carry out calculations involving energy transfer.

Newton's third law

Newton's third law states that for every action there is an equal but opposite reaction.

Action and reaction refer to forces, just as push and pull are forces.

The law means that when A pushes B, B pushes A back with the same size of force. This is called an interaction and involves two objects.

Examples of Newton's third law are:

● When you push against a wall – you push on the wall (the action), the wall pushes back on you (the reaction) with the same size of force.
● When you swim in a swimming pool – you push the water backwards (the action), the water pushes you forwards (the reaction) with the same size of force.
● When you walk across the room – you push the floor backwards (the action), the floor pushes you forwards (the reaction) with the same size of force.
● When you bang your fist on the table – your fist exerts a force on the table (the action), the table exerts a force on your fist (the reaction) with the same size of force.
● When you stand up – you push the floor downwards (the action), the floor pushes you upwards (the reaction) with the same size of force.

the wall pushes you (reaction)

you push the wall (action)

Figure 11.1

● When you release a blown up balloon – the balloon pushes the air inside the balloon backwards (the action), the air pushes the balloon forwards (the reaction) with the same size of force.

In Newton's third law the action and the reaction are equal in size but opposite in direction. The action acts on one object while the reaction acts on a different object – the two objects interact. Although the forces are equal in size and act in opposite directions, this is not the same as Newton's first law since the forces act on *different* objects.

Figure 11.2

Figure 11.4 Hot exhaust gases from a Typhoon

outer space there is a vacuum (no air and therefore no oxygen) so space rockets must carry their own supply of oxygen as well as the fuel for burning.

However, rockets and jet aircraft do not need air to push against in order to move. The rocket or jet aircraft moves because of the interaction between the vehicle and the exhaust gases.

When a rocket lifts off, there is an interaction between the rocket and the exhaust gases. The rocket pushes the exhaust gases down (the action). The exhaust gases push the rocket up (the reaction). If this reaction is greater than the weight of the rocket, there is an unbalanced force acting upwards on the rocket and the rocket accelerates upwards.

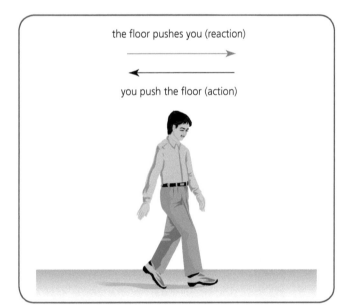

Figure 11.3

The motion of rockets and jet aircraft can be explained using Newton's third law. In both cases a high speed stream of hot gases, produced by burning fuel, is pushed backwards from the rocket or jet engine with a large force. From Newton's third law, a force of the same size pushes the vehicle forwards:

- Action – vehicle pushes hot gases backwards.
- Reaction – hot gases push vehicle forwards with the same size of force.

For the fuel to burn and produce hot gases (called exhaust gases) a supply of oxygen is needed. Jet aircraft use the oxygen in the surrounding air. However, in

Figure 11.5 A space shuttle just after lift-off

Work and energy

Energy is an important and useful physical quantity. Energy has different forms and although it is difficult to define energy, there are ways we can use to measure the different forms.

- Energy gets things done and lets you do things – we call this work.
- When work is done, energy is transferred (to an object or into other forms of energy).

Consider an object of mass m being pushed along a horizontal bench, as in Figure 11.6.

Figure 11.6 Work is done in pushing an object through a distance

work done on object = energy transferred

work done = force applied × distance moved by the force

$$E_W = F \times d$$

where E_W = work done, measured in joules (J),
F = force applied, measured in newtons (N),
d = distance moved by the force, measured in metres (m).

One joule is the work done when a force of 1 newton moves a distance of 1 metre.

Therefore 1 joule = 1 newton metre ($1\,J = 1\,Nm$).

Worked examples

Example 1

A supermarket trolley is pushed at constant speed by a horizontal force of 5.4 N. The trolley travels a distance of 20 m. Calculate the work done by this force.

Solution

$E_W = F \times d$

$E_W = 5.4 \times 20$

$E_W = 108\,J$

Example 2

A girl is pedalling her bicycle along a horizontal road. She exerts a constant force of 15 N on the pedals. The girl and bicycle gain 1800 J of energy. Calculate the maximum distance travelled by the girl and her bicycle.

Solution

$E_W = F \times d$

$1800 = 15 \times d$

$d = \dfrac{1800}{15} = 120\,m$

Example 3

The mass of a table is 35 kg. A person pushes the table 750 mm horizontally so that it is against a wall. To move the table this distance, 65 J of work are done. Calculate the horizontal force the person exerts on the table.

Solution

Note: $750\,mm = 750 \times 10^{-3}\,m = 0.750\,m$

$E_W = F \times d$

$65 = F \times 750 \times 10^{-3}$

$F = \dfrac{65}{750 \times 10^{-3}} = 86.7 = 87\,N$

Gravitational potential energy

When an object is raised from the floor onto a shelf, work has to be done to overcome the force of gravity that acts on the object. As a result, energy has been transferred to the object owing to its position now being further above the centre of the Earth. The object is said to have gained **gravitational potential energy** (E_p).

If the object were to fall from the shelf then this gravitational potential energy would be changed into other forms of energy such as kinetic energy, heat and perhaps sound.

Consider an object of mass m moving upwards at constant speed through a height h (Figure 11.7).

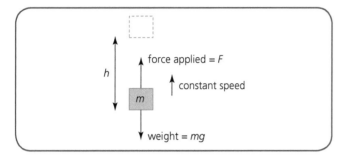

Figure 11.7 The work done in lifting an object at constant speed is transferred into a gain in gravitational potential energy

The downward force acting on the object is its weight (= mg). Since the object is moving at constant speed, balanced forces act on the object (Newton's first law). This means that there must be a steady upwards force of mg applied to the object. The effects of air resistance can be ignored.

work done = energy transferred to object

work done = force applied × distance moved by the force

work done = $F \times d$ but $F = mg$ and $d = h$, so

work done = mgh

work done = change in gravitational potential energy (E_p)

change in gravitational potential energy	=	mass of object	×	gravitational field strength	×	change in height

$$E_p = mgh$$

where E_p = change (gain or loss) in gravitational potential energy, measured in joules (J),
m = mass of object, measured in kilograms (kg),
g = gravitational field strength, measured in newtons per kilogram (N kg^{-1}) – for the Earth g is 9.8 N kg^{-1},
h = change in height, measured in metres (m).

A table of gravitational field strengths is given on the *Data Sheet* on page 170.

Worked examples

Example 1

The mass of a bag of sand is 25 kg. The bag is lifted 0.70 m vertically from the ground onto the back of a lorry. Calculate the gravitational potential energy gained by the bag of sand.

Solution

gain in $E_p = mgh$
gain in $E_p = 25 \times 9.8 \times 0.7$
gain in $E_p = 172$ J

Example 2

A can of soup falls from a kitchen cupboard onto the floor. The can falls through a height of 0.60 m. The can loses 2.4 J of gravitational potential energy. Calculate the mass of the can.

Solution

loss in $E_p = mgh$
$2.4 = m \times 9.8 \times 0.6$
$m = \dfrac{2.4}{5.88} = 0.41$ kg

Example 3

The mass of a diver is 40 kg. The diver steps off a diving board and drops towards the water below. When she just enters the water, the loss in gravitational potential energy of the diver is 2.0 kJ. Calculate the height of the diving board above the water.

Solution

Note: 2.0 kJ = 2 × 10³ J = 2000 J
loss in $E_p = mgh$
$2000 = 40 \times 9.8 \times h$
$h = \dfrac{2000}{392} = 5.1$ m

Kinetic energy

Kinetic energy (E_K) is the energy associated with any moving object.

Consider an object of mass m starting from rest ($u = 0$). The object is accelerated in a straight line by a constant unbalanced force F_{un}. The acceleration of the object is a. After a time t, the speed of the object is v. The effects of air resistance can be ignored.

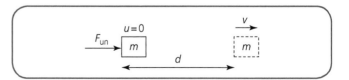

Figure 11.8 The work done in accelerating an object from rest is transferred into a gain in kinetic energy

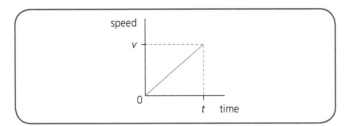

Figure 11.9 Speed–time graph for an object accelerated from rest

work done = energy transferred to object

work done = force applied × distance moved by the force

work done = $F_{un} \times d$

$F_{un} = ma$ and d = area under speed–time graph, so

work done = ma × area under graph

$a = \dfrac{v-u}{t} = \dfrac{v-0}{t} = \dfrac{v}{t}$ and area under graph = $\dfrac{1}{2}vt$, so

work done = $\dfrac{mv}{t} \times \dfrac{1}{2}vt$

work done = $\dfrac{1}{2}mv^2$

work done = gain in kinetic energy = $\dfrac{1}{2}mv^2$

For there to be a change (gain or loss) in kinetic energy, there must be a difference between the initial and final kinetic energies of the object (one is larger than the other). The difference is always the large kinetic energy minus the small kinetic energy:

$$\begin{matrix}\text{change in} \\ \text{kinetic energy}\end{matrix} = \begin{matrix}\text{kinetic energy} \\ \text{(large)}\end{matrix} - \begin{matrix}\text{kinetic energy} \\ \text{(small)}\end{matrix}$$

In the above case, the small kinetic energy of the object is zero, since the object started from rest.

$$\begin{matrix}\text{gain in} \\ \text{kinetic energy}\end{matrix} = \begin{matrix}\text{kinetic energy} \\ \text{(large)}\end{matrix} - \begin{matrix}\text{kinetic energy} \\ \text{(small)}\end{matrix}$$

$\dfrac{1}{2}mv^2$ = kinetic energy (large) − 0

i.e. kinetic energy (large) = $\dfrac{1}{2}mv^2$ = kinetic energy of object

$$E_K = \dfrac{1}{2}mv^2$$

where E_K = kinetic energy of object, measured in joules (J), m = mass of the object, measured in kilograms (kg), v = speed of the object, measured in metres per second (m s^{-1}).

Above we considered an object starting from rest ($u = 0$ at $t = 0$). However, what if the object is already moving with a speed, u?

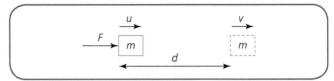

Figure 11.10

When the object has a positive acceleration ($v > u$) then:

$$\text{change (gain) in } E_K = \dfrac{1}{2}mv^2 - \dfrac{1}{2}mu^2$$
$$= E_K(\text{large}) - E_K(\text{small})$$

When the object has a negative acceleration ($u > v$) then:

$$\text{change (loss) in } E_K = \dfrac{1}{2}mu^2 - \dfrac{1}{2}mv^2$$
$$= E_K(\text{large}) - E_K(\text{small})$$

In general:

$$\text{change (gain or loss) in } E_K = E_K(\text{large}) - E_K(\text{small})$$

Kinetic energy of an object, $E_K = \dfrac{1}{2}mv^2$. If the mass of an object is constant, then E_K depends on v^2:

- When the speed of the object doubles (× 2) then the kinetic energy is four times greater (× $2^2 = 4$).
- When the speed of the object trebles (× 3), the kinetic energy is nine times greater (× $3^2 = 9$).
- When the speed of the object quadruples (× 4), the kinetic energy is sixteen times greater (× $4^2 = 16$).

Worked examples

Example 1

The mass of an object is 40 kg. The speed of the object is 8.0 m s^{-1}. Calculate the kinetic energy of the object.

Solution

$$E_K = \dfrac{1}{2}mv^2$$

$$E_K = \dfrac{1}{2} \times 40 \times (8)^2$$

$$E_K = 1280\,\text{J}$$

Example 2

A car is travelling along a straight road at a speed of $14\,\mathrm{m\,s^{-1}}$. When travelling at this speed, the kinetic energy of the car is $78.4\,\mathrm{kJ}$. Calculate the mass of the car.

Solution

Note: $78.4\,\mathrm{kJ} = 78.4 \times 10^3\,\mathrm{J} = 78\,400\,\mathrm{J}$

$$E_K = \frac{1}{2}mv^2$$

$$78.4 \times 10^3 = \frac{1}{2} \times m \times 14^2$$

$$78.4 \times 10^3 = \frac{1}{2}m \times 196$$

$$m = \frac{78.4 \times 10^3}{98} = 800\,\mathrm{kg}$$

Example 3

An archer fires an arrow from a bow. The mass of the arrow is $20\,\mathrm{g}$. The kinetic energy of the arrow as it leaves the bow is $36\,\mathrm{J}$. Calculate the speed of the arrow as it leaves the bow.

Solution

Note: $20\,\mathrm{g} = 20 \times 10^{-3}\,\mathrm{kg} = 0.020\,\mathrm{kg}$

$$E_K = \frac{1}{2}mv^2$$

$$36 = \frac{1}{2} \times 20 \times 10^{-3} \times v^2$$

$$v^2 = \frac{36}{10 \times 10^{-3}} = 3600$$

$$v = \sqrt{3600} = 60\,\mathrm{m\,s^{-1}}$$

Example 4

The mass of a car is $1200\,\mathrm{kg}$. The car slows down from $22\,\mathrm{m\,s^{-1}}$ to $10\,\mathrm{m\,s^{-1}}$. Calculate the change in kinetic energy of the car.

Solution

$$E_K\,(large) = \frac{1}{2}mv^2 = \frac{1}{2} \times 1200 \times 22^2 = 290\,400\,\mathrm{J}$$

$$E_K\,(small) = \frac{1}{2}mv^2 = \frac{1}{2} \times 1200 \times 10^2 = 60\,000\,\mathrm{J}$$

$$\text{change in } E_K = E_K\,(large) - E_K\,(small)$$

$$= 290\,400 - 60\,000$$

$$= 230\,400\,\mathrm{J}$$

$$= 2.30 \times 10^5\,\mathrm{J}$$

Conservation of energy

Energy cannot be created or destroyed but it can be transferred from one form into another when work is done. When a system 'gains' energy then another system must have 'lost' the same amount of energy; an energy transfer has taken place.

When work is done there is a transfer of energy.

Energy transfer or energy change

When work is done on a system, energy is transferred to or from that system:

- When an object on a horizontal surface (so no change in gravitational potential energy) is acted on by an unbalanced force then the speed and kinetic energy of the object will change. From the law of conservation of energy:

$$\frac{\text{work done}}{\text{by unbalanced force}} = \frac{\text{change in}}{\text{kinetic energy}}$$

$$\begin{array}{c}\text{unbalanced}\\\text{force}\end{array} \times \begin{array}{c}\text{distance moved}\\\text{by force}\end{array} = E_K(\text{large}) - E_K(\text{small})$$

- When an object is raised vertically at a constant speed (so no change in kinetic energy), the gravitational potential energy of the object will change. From the law of conservation of energy:

$$\text{work done} = \text{change in gravitational potential energy}$$

$$\text{force applied} \times \text{distance moved by force} = mgh$$

During any energy transformation, the total amount of energy is always conserved, but may be changed into less useful forms.

Example of an energy transfer: a falling object

As an object falls freely from rest (initial velocity $u = 0$), it 'loses' gravitational potential energy but 'gains' kinetic energy. Just before impact with the ground it has a final velocity v, having 'lost' all of its gravitational potential energy. The object now has only kinetic energy.

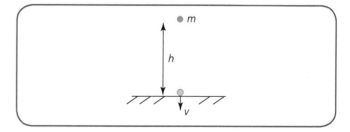

Figure 11.11

When a constant resistive force (F_r) due to air resistance is present, the loss in gravitational potential energy is transferred into a gain in kinetic energy *and* work done against the resistive force (in the form of heat and perhaps sound, which are 'lost' or dissipated from the system). Then:

loss in gravitational E_P = gain in E_K + work done against the resistive force

$$mgh = \left(\frac{1}{2}mv^2 - 0\right) + (F_r \times d)$$

$d = h$, so

$$mgh = \frac{1}{2}mv^2 + (F_r \times h)$$

When there is no resistive force there is no work done against the resistive force and there is no energy 'lost' (transferred) to the surroundings:

loss in gravitational E_P = gain in E_K

$$mgh = \frac{1}{2}mv^2 - 0$$

$$mgh = \frac{1}{2}mv^2$$

When an object is projected vertically upwards with an initial velocity u, it slows down and so 'loses' kinetic energy and in gaining height 'gains' gravitational potential energy. The object will come to rest (final velocity $v = 0$) when it is at its maximum height (h) above the ground.

When a constant resistive force (F_r) due to air resistance is present, the loss in kinetic energy is transferred into a gain in gravitational potential energy *and* work done against the resistive force (in the form of heat and perhaps sound, which are 'lost' or dissipated from the system). Then:

loss in E_K = gain in gravitational E_P + work done against the resistive force

$$\frac{1}{2}mu^2 - 0 = mgh + (F_r \times d)$$

$d = h$, so

$$\frac{1}{2}mu^2 = mgh + (F_r \times h)$$

When there is no resistive force then there is no work done against the resistive force and there is no energy 'lost' (transferred) to the surroundings:

loss in E_K = gain in gravitational E_P

$$\frac{1}{2}mu^2 - 0 = mgh$$

$$\frac{1}{2}mu^2 = mgh$$

Worked examples

Example

A ball is released from the window of a house so that it falls onto the ground below. The mass of the ball is 250 g. The ball is travelling at $10\,\text{m s}^{-1}$ just as it is about to hit the ground. The effects of air resistance can be ignored.

a) Calculate the gain in kinetic energy of the ball.

b) Calculate the height from which the ball was dropped.

Solution

a) Note: initial $E_K = 0 = E_K$(small)
   Note: $250\,\text{g} = 250 \times 10^{-3}\,\text{kg} = 0.250\,\text{kg}$

   gain in $E_K = \frac{1}{2}mv^2 - 0$

   gain in $E_K = \frac{1}{2}mv^2$

   gain in $E_K = \frac{1}{2} \times 250 \times 10^{-3} \times 10^2$

   gain in $E_K = 12.5\,\text{J}$

b) No energy is transferred with the surroundings (since the effects of air resistance can be ignored and so no work is done by air resistance). So:

   loss in gravitational E_P = gain in E_K

   $$mgh = 12.5$$

   $250 \times 10^{-3} \times 9.8 \times h = 12.5$ Note: g from *Data Sheet*

   $$h = \frac{12.5}{2.45} = 5.1\,\text{m}$$

Example of an energy transfer: a slope

An object is placed on a slope.

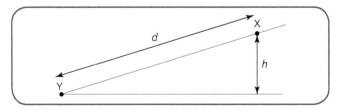

Figure 11.12

The object is placed initially at X (initial velocity $u = 0$). The object accelerates down the slope. The object 'loses' gravitational potential energy but 'gains' kinetic energy. At Y it has a velocity v.

When a constant frictional force (F_r) is present, the loss in gravitational potential energy is transferred into a gain in kinetic energy *and* work done against the frictional force (in the form of heat and perhaps sound, which are 'lost' or dissipated from the system). So:

$$\text{loss in gravitational } E_P = \text{gain in } E_K + \text{work done against friction}$$

$$mgh = \left(\frac{1}{2}mv^2 - 0\right) + (F_r \times d)$$

$$mgh = \frac{1}{2}mv^2 + (F_r \times d)$$

When there is no frictional force then there is no work done against friction and there is no energy 'lost' (transferred) to the surroundings:

$$\text{loss in gravitational } E_P = \text{gain in } E_K$$

$$mgh = \frac{1}{2}mv^2 - 0$$

$$mgh = \frac{1}{2}mv^2$$

When an object at Y is projected up the slope with an initial velocity u, it slows down and so 'loses' kinetic energy and in gaining height 'gains' gravitational potential energy. The object will come to rest at X (final velocity $v = 0$) when it is at its maximum height (h) up the slope.

When a constant frictional force (F_r) is present on the slope, the loss in kinetic energy is transferred into a gain in gravitational potential energy *and* work done against the frictional force (in the form of heat and perhaps sound, which are 'lost' or dissipated from the system). So:

$$\text{loss in } E_K = \text{gain in gravitational } E_P + \text{work done against friction}$$

$$\frac{1}{2}mu^2 - 0 = mgh + (F_r \times d)$$

$$\frac{1}{2}mu^2 = mgh + (F_r \times d)$$

When there is no frictional force then there is no work done against friction and there is no energy 'lost' (transferred) to the surroundings:

$$\text{loss in } E_K = \text{gain in gravitational } E_P$$

$$\frac{1}{2}mu^2 - 0 = mgh$$

$$\frac{1}{2}mu^2 = mgh$$

Worked example

Example

A boy on a skateboard starts from rest at point X on a slope. He accelerates down the slope to point Y (Figure 11.13).

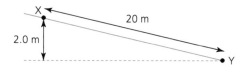

Figure 11.13

Point X is 2.0 m vertically above point Y. Points X and Y are 20 m apart. The mass of the boy and skateboard is 60 kg.
a) Calculate the change in gravitational potential energy of the boy and skateboard as he moves from X to Y.
b) A constant frictional force of 8.0 N, parallel to the slope, acts on the boy and skateboard as he moves from X to Y. Calculate the speed of the boy and skateboard at Y.

Solution

a) Note: g from *Data Sheet*
change in gravitational $E_P = mgh = 60 \times 9.8 \times 2$
change in gravitational $E_P = 1176\,J$
(= loss in gravitational E_P)

b) Note: initial $E_K = 0 = E_K$(small)

$$\text{loss in gravitational } E_P = \text{gain in } E_K + \text{work done against friction}$$

$$1176 = \left(\frac{1}{2}mv^2 - 0\right) + (F_r \times d)$$

$$1176 = \left(\frac{1}{2} \times 60 \times v^2\right) + (8 \times 20)$$

$$1176 = 30v^2 + 160$$

$$30v^2 = 1176 - 160$$

$$v^2 = \frac{1016}{30} = 33.9$$

$$v = \sqrt{33.9} = 5.8\,m\,s^{-1}$$

Example of an energy transfer: pendulum

When a pendulum bob, shown in Figure 11.14, is moved from point B to point A, work is done and energy is transferred to the bob.

$$\text{work done on pendulum bob} = \text{energy transferred to bob}$$
$$= \text{gain in gravitational potential energy}$$

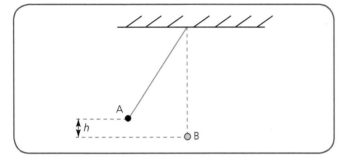

Figure 11.14

The bob is now released from A. The bob is at rest initially (initial velocity $u = 0$). As the bob swings towards B it 'loses' gravitational potential energy and 'gains' kinetic energy. At B it has a (final) velocity v.

When a constant resistive force (F_r) is present, the loss in gravitational potential energy is transferred into a gain in kinetic energy *and* work done against the resistive force (in the form of heat and perhaps sound, which are 'lost' or dissipated from the system). So:

$$\text{loss in gravitational } E_P = \text{gain in } E_K + \text{work done against the resistive force}$$

$$mgh = \left(\frac{1}{2}mv^2 - 0\right) + (F_r \times d)$$

where d is the arc of the circle from A to B.

$$mgh = \frac{1}{2}mv^2 + (F_r \times d)$$

When there is no resistive force then there is no work done against the resistive force and there is no energy 'lost' (transferred) to the surroundings:

$$\text{loss in gravitational } E_P = \text{gain in } E_K$$

$$mgh = \frac{1}{2}mv^2 - 0$$

$$mgh = \frac{1}{2}mv^2$$

When the bob swings from A to B it 'loses' gravitational potential energy but 'gains' kinetic energy. At B it has maximum speed and hence maximum kinetic energy.

Worked example

Example

Figure 11.15 shows a pendulum bob at position X, its rest position. The mass of the pendulum bob is 30 g.

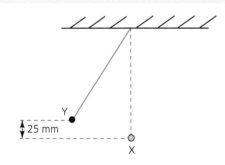

Figure 11.15

The pendulum bob is pulled to point Y. Y is 25 mm above the rest position, X.

a) When the pendulum bob is moved from X to Y, calculate the gain in gravitational potential energy of the bob.

b) The pendulum bob is released from Y and swings to and fro until it comes to rest.
   i) Describe the energy changes that take place as the pendulum bob swings from Y to X.
   ii) Show that the maximum possible speed of the pendulum bob at X is 0.70 m s⁻¹. State any assumption you make in your calculation.

Solution

a) Note: 30 g = 30 × 10⁻³ kg = 0.030 kg
   and 25 mm = 25 × 10⁻³ m = 0.025 m

   gain in gravitational potential energy $E_P = mgh$
   gain in E_P = 30 × 10⁻³ × 9.8 × 25 × 10⁻³
   gain in E_P = 7.4 × 10⁻³ J

b) i) Since the pendulum bob eventually comes to rest there has to be some work done against resistive forces. Therefore, as the bob swings from Y to X, gravitational potential energy is changed into kinetic energy and some heat (due to work done against the resistive force).

ii) Assuming that no energy is transferred with the surroundings:

gain in E_K = loss in gravitational E_P

$$\frac{1}{2}mv^2 - 0 = mgh \quad \text{Note: initial } E_K = 0 = E_K(\text{small})$$

$$\frac{1}{2}mv^2 = 7.4 \times 10^{-3}$$

$$\frac{1}{2} \times 30 \times 10^{-3} \times v^2 = 7.4 \times 10^{-3}$$

$$v^2 = \frac{7.4 \times 10^{-3}}{15 \times 10^{-3}} = 0.49$$

$$v = \sqrt{0.49} = 0.70 \, \text{m s}^{-1}$$

Power

Power is the energy transferred (or work done) in 1 second.

$$\text{power} = \frac{\text{energy transferred}}{\text{time taken}} = \frac{\text{work done}}{\text{time taken}}$$

$$P = \frac{E}{t}$$

where P = power, measured in watts (W),
E = energy transferred (or work done), measured in joules (J),
t = time taken, measured in seconds (s).

The unit used to measure power is the watt (W); 1 watt means that 1 joule of energy has been transferred (or 1 joule of work has been done) in 1 second.

Therefore, 1 watt is 1 joule per second ($1 \, \text{W} = 1 \, \text{J s}^{-1}$)

Sometimes, although work is done on an object, the object continues to move *at a constant speed* in a straight line. In this case:

$$\text{power} = \frac{\text{work done}}{\text{time taken}}$$

$$\text{power} = \frac{\text{force applied} \times \text{distance moved by force}}{\text{time taken}}$$

$$\text{average speed} = \frac{\text{distance moved by force}}{\text{time taken}}$$

So

power = force applied × average speed

Worked example

Example

A lift in a building travels vertically upwards at constant speed. The lift rises through a height of 20 m in 40 s. The mass of the lift is 1200 kg.
a) Calculate the gravitational potential energy gained by the lift.
b) Calculate the minimum power needed by the lift motor.

Solution

a) $E_P = mgh$

$E_P = 1200 \times 9.8 \times 20$ Note: g from *Data Sheet*

$E_P = 235\,200 \, \text{J}$

b) $\text{power} = \dfrac{\text{work done on lift}}{\text{time taken}} = \dfrac{\text{gain in } E_P \text{ of lift}}{\text{time taken}}$

$$P = \frac{E_P}{t} = \frac{mgh}{t} = \frac{235\,200}{40} = 5880 \, \text{W}$$

or

As the lift is travelling at constant speed, the forces on the lift are balanced (Newton's first law):

force applied (upwards)   =   weight (downwards)
to lift, F of lift, mg

$$\text{average speed} = \frac{\text{distance travelled}}{\text{time taken}} = \frac{20}{40} = 0.5 \, \text{m s}^{-1}$$

$$\text{power} = \frac{\text{work done on lift}}{\text{time taken}}$$

$$P = \frac{F \times d}{t} = mg \times \text{average speed}$$

$$P = 1200 \times 9.8 \times 0.5 = 5880 \, \text{W}$$

Key facts and physics equations: Newton's third law and energy

- Newton's third law – 'for every action there is an equal but opposite reaction'.
- In Newton's third law, the action and reaction forces are equal in size but act on different objects.
- Work done = force applied × distance moved by the force, i.e. $E_W = F \times d$.
- Work done is measured in joules (J), force applied in newtons (N) and distance moved by the force in metres (m).
- Change in gravitational potential energy $E_p = mgh$.

- Gravitational potential energy (gain or loss) is measured in joules (J), mass in kilograms (kg), gravitational field strength in newtons per kilogram ($N\,kg^{-1}$) and height in metres (m).
- Kinetic energy $E_K = \frac{1}{2}mv^2$.
- Kinetic energy is measured in joules (J), mass in kilograms (kg) and speed in metres per second ($m\,s^{-1}$).
- Change in kinetic energy = E_K(large) − E_K(small).
- Power = $\frac{\text{work done}}{\text{time}}$, i.e. $P = \frac{E_W}{t}$.
- Power is measured in watts (W), work done in joules (J) and distance in metres (m).
- Energy can be transferred from one form into another.

End-of-chapter questions

Information, if required, for use in the following questions can be found on the *Data Sheet* on page 170.

1 In the following sentences the words represented by the letters A, B and C are missing.

'When a rocket lifts-off from a planet, the action force is the rocket pushing the _____A_____ down. The reaction force is the exhaust gases pushing the _____B_____ up. The reaction causes the rocket to lift off from the planet.' This is an example of Newton's _____C_____ law.

Match each letter with the correct word or words from the following list:

exhaust gases; first; ground; rocket; second; third

2 While cutting his lawn, a man pushes a lawnmower with a horizontal force of 35 N. The lawn is 4.0 m wide. The man has to cross the lawn 16 times in order to cut all of the grass on the lawn. Calculate the minimum work done by the man in cutting the lawn.

3 A boy pushes a go-kart, at constant speed, with a horizontal force of 25 N. The boy does 500 J of work in pushing the go-kart. Calculate the maximum distance the go-kart moves through.

4 An obstruction is moved horizontally through a distance of 100 m. This requires 1.5 MJ of work to be done. Calculate the horizontal force exerted on the obstruction.

5 A box of cereal is taken from a cupboard and placed on a table. The table is 350 mm vertically below the cupboard. The mass of the box is 750 g. Calculate the change in gravitational potential energy of the box.

6 The mass of a pallet of bricks is 350 kg. The pallet is lifted by a crane from the ground onto the back of a lorry. The pallet of bricks gains 2450 J of gravitational potential energy. Calculate the height of the back of the lorry above the ground.

7 A lift in a building rises through a vertical height of 8.0 m. During an ascent the lift gains 96 000 J of gravitational potential energy. Calculate the mass of the lift.

8 The mass of the ocean liner *Queen Mary II* is 150×10^6 kg. The *Queen Mary II* has a cruising speed of $13\,m\,s^{-1}$ and a maximum speed of $16\,m\,s^{-1}$.
   a) Calculate the kinetic energy of the *Queen Mary II* while travelling at its cruising speed.
   b) The *Queen Mary II* speeds up from its cruising speed to its maximum speed. Calculate the change in kinetic energy of the liner.

9 The kinetic energy of a satellite in orbit above the Earth is 2.93×10^{10} J. The satellite is travelling at a speed of $3100\,m\,s^{-1}$. Calculate the mass of the satellite.

10 A car is travelling along a road. The kinetic energy of the car is 260 kJ. The mass of the car is 1000 kg. Calculate the speed of the car.

11 Mary does an experiment to estimate her power in running up stairs. She records the following information:

height of one step = 200 mm; number of steps = 30; time taken to run up 30 steps = 6.0 s; mass of Mary = 45 kg
   a) Calculate the vertical height Mary moved through.
   b) Calculate the gravitational potential energy gained by Mary in running up the stairs.
   c) Calculate an estimate for Mary's power in running up the stairs.

d) Explain whether your answer is an overestimate or an underestimate of the power Mary would use to run up the stairs.

12 A boy goes sledging during the winter. The combined mass of the boy and sledge is 60 kg. The boy takes his sledge from the bottom to the top of a slope. The height of the slope is 14 m, as shown in Figure 11.16.

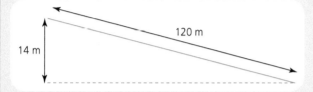

120 m

14 m

Figure 11.16

a) Calculate the gravitational potential energy gained by the boy and his sledge in going from the bottom to the top of the slope.

b) The boy and his sledge then start from rest at the top of the slope and descend to the bottom. A constant frictional force of 34 N, parallel to the slope, acts on the sledge during the descent.

i) Calculate the work done against friction during the descent.

ii) Show that the speed of the boy and sledge at the bottom of the slope is $11.8\,\text{m s}^{-1}$.

13 At an indoor rock-climbing centre, a girl climbs a rock face (Figure 11.17). The mass of the girl is 50 kg. The girl is attached to a rope. She climbs through a vertical height of 3.0 m from the ground. The time she takes to do this is 15 s.

a) i) Calculate the gravitational potential energy gained by the girl in climbing 3.0 m.

ii) Calculate the minimum power required for the girl to climb the 3.0 m.

b) A safety device is fitted to the rope attached to the girl. At the top of her climb the girl falls vertically off the rock face. The safety device ensures that the girl falls safely to the ground. Just as she reaches the ground, the speed of the girl is $1.4\,\text{m s}^{-1}$. Calculate the average frictional force exerted by the safety device on the girl.

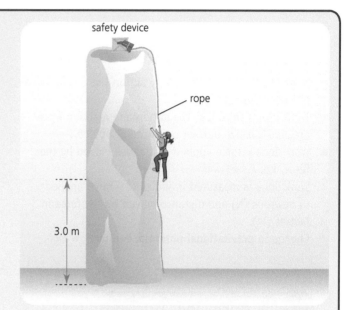

safety device

rope

3.0 m

Figure 11.17

14 A skier is travelling down a slope (Figure 11.18). The mass of the skier is 70 kg. The speed of the skier at point X is $4.0\,\text{m s}^{-1}$. The speed of the skier at point Y is $15\,\text{m s}^{-1}$. Calculate the average frictional force, parallel to the slope, acting on the skier as the skier accelerates from X to Y.

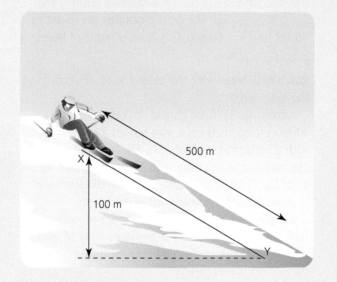

X

500 m

100 m

Y

Figure 11.18

15 A pendulum bob, initially at rest, is released and swings from point A to point B as shown in Figure 11.19. The mass of the bob is 0.3 kg. The height of A above B is 40 mm.

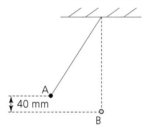

A

40 mm

B

Figure 11.19

a) Calculate the loss in gravitational potential energy of the bob as it swings from A to B.

b) Assuming all the gravitational potential energy at A is changed into kinetic energy at B, calculate the maximum speed of the bob at B.

16 The mass of a bucket is 150 kg. A crane lifts the bucket through a vertical height of 25 m to the top of a building. The time taken to do this is 40 s.

a) Calculate the gravitational potential energy gained by the bucket.

b) Calculate the minimum power of the crane's motor.

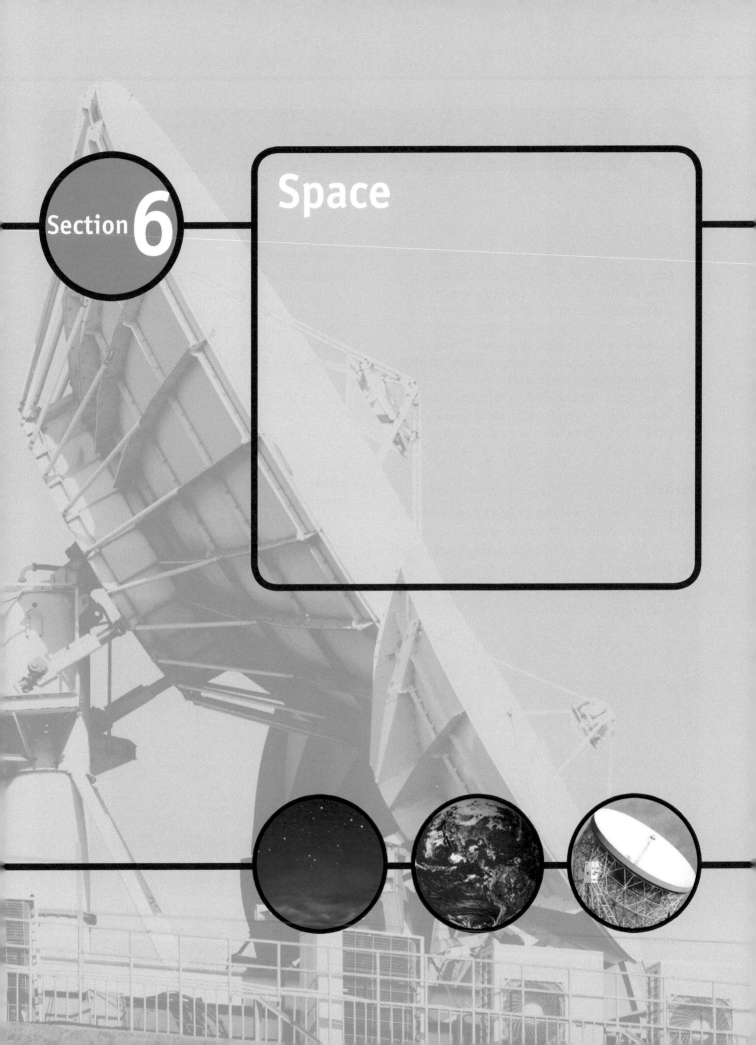

Section **6**

Space

Space exploration

Learning outcomes

At the end of this chapter you should be able to:

1 Use the following terms correctly in context: asteroid, planet, moon, dwarf planet, sun, star, solar system, exoplanet, galaxy, universe.
2 State a benefit of using satellites such as: GPS, weather forecasting, communications, scientific discovery, space exploration.
3 Describe a risk associated with space travel such as: ionisating radiations, pressure differential, extreme temperatures, take-off, re-entry.
4 Explain what happens to a spacecraft when it re-enters the atmosphere.
5 State that the period of a satellite's orbit depends on its altitude above the Earth.
6 State that a geostationary satellite at an altitude of 36 000 km has a period of 24 hours.

Space

Humankind has for centuries looked up and wondered at the vastness of the night sky, our window on the universe. We have long dreamt of visiting other planets and distant worlds.

Figure 12.1 The night sky

This chapter will introduce a number of useful terms including:

● asteroid – a relatively small, inactive lump of rock that orbits the Sun

● planet – an object that orbits a star
● moon – an object that orbits a planet
● star – a ball of very hot gas produced by nuclear fusion
● galaxy – an immense system of stars, dust and gas
● universe – the whole of space; everything!

Figure 12.2 The Earth from space

We live on a planet called **Earth**. The Earth is the third of eight planets that orbit around the **Sun**. The Sun is a star that produces enormous amounts of heat and light. The Sun and the eight planets form our **solar system** (see Figures 12.3 and 12.4).

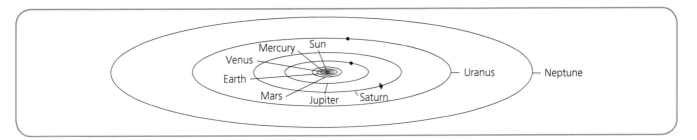

Figure 12.3 Our solar system

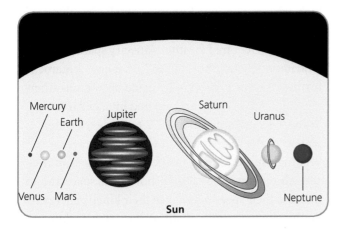

Figure 12.4 Relative sizes of our Sun and its planets

There are many other objects that revolve around the Sun, such as dwarf planets. The main difference between a planet and a dwarf planet is that a planet has cleared the path around the Sun while a dwarf planet orbits along with similar objects, such as asteroids, that can cross their path around the Sun.

The solar system is a small part of thousands of millions of stars that form part of a **galaxy** called the **Milky Way** (Figure 12.5). In turn there are millions more galaxies, which together form the **universe**.

Figure 12.5 The Milky Way

Stars outside our solar system, just like our Sun, have planets orbiting around them. However, as these planets are so far away from the Earth, we are unable to see them and they are called extrasolar planets or **exoplanets**.

Space triumphs

In the thirteenth century the Chinese used crude fireworks filled with gunpowder against their enemies. In the twentieth century, rockets evolved into sophisticated multi-stage systems fuelled by supercooled combustible liquids. After the Second World War a 'space race' developed between America and Russia. Russia initially triumphed over America by being the first to put a satellite, 'Sputnik', into orbit above the Earth in 1957, first to put a human, Yuri Gagarin, in space in 1961 and the first to soft-land a robotic craft on the Moon in 1966. However, the Americans went on to win the 'space prize' of putting the first humans, Neil Armstrong and 'Buzz' Aldrin, on the Moon on the *Apollo 11* flight. On placing his left foot on the Moon, Neil Armstrong declared 'that's one small step for man, one giant leap for mankind'. The Americans were also the first to develop a reusable space vehicle, the 'space shuttle'.

Space disasters

Space is not without danger. A number of astronauts, both American and Russian, have died during space missions. Two of the most recent disasters involved the space shuttles *Challenger* and *Columbia*. The *Challenger* (1986) disaster occurred 73 seconds after lift-off as a result of a failure of a component on one of the external solid rocket boosters. All seven crew members, one of whom was a school teacher, died. The *Columbia* (2003) disintegrated during re-entry into the Earth's atmosphere as a result of damage to the thermal tiles on the left wing. Again all seven crew were killed.

Figure 12.6 A Saturn V rocket being used to launch an Apollo spacecraft

Challenges of space travel

The vast distances involved in space travel mean that a space mission can take years to complete. To shorten this time, spacecraft have to travel at very high speeds. This can be achieved by:

- Accelerating the spacecraft to very high speeds using the force (thrust) of the rocket engine.
  However, to launch a spacecraft from the Earth requires a huge amount of fuel (and therefore mass), which is burned to get the spacecraft into Earth orbit – approximately 90 per cent of the initial launch mass of a spacecraft is fuel. Burning the small amount of fuel left produces a large acceleration but only for a short time and this does not produce the very high speeds required. To overcome this, scientists have developed new methods to propel spacecraft when in space. Any object that propels material in one direction will be propelled in the opposite direction – Newton's

third law. That material does not need to be burning, but it does need to produce a force. Three types of electric (ion) rocket engine have been tested in space: electrothermal, using electric elements to heat up a propellant; electrostatic, which ionises the propellant and then accelerates it through an electric field; and electrodynamic systems, which generate a plasma (very hot charged particles) and then accelerate it with an electric or magnetic field. Each one of these can provide high exhaust velocities, but only a small force that could be used to accelerate the spacecraft over time to the very high speeds required for interplanetary space flight.

- Gravitational 'slingshot'. In 2012, Voyager 1 became the furthest travelled object to be made by humans when it left our solar system and entered interstellar space. Voyager 1 was launched in 1977 to study the outer planets of the solar system. It was able to travel at very high speeds due to an effect called **gravitational slingshot**. Gravitational slingshot allows spacecraft to increase their speed and hence their kinetic energy. To do this the spacecraft's flight path must pass close to a planet. With the correct trajectory, the spacecraft uses the gravitational pull of the planet to increase its speed and swing it towards its target (in the case of Voyager 1, another outer planet).

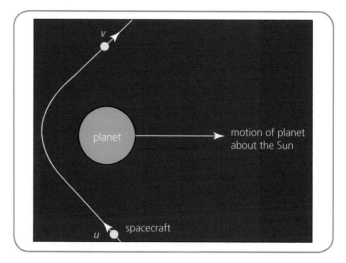

Figure 12.7 Gravitational slingshot about a planet

The slingshot is needed so that the spacecraft gains enough speed to overcome the gravitational pull of the Sun on the spacecraft and so allow it to reach other distant planets.

Challenges of manned space exploration

Space is a very hostile place for humans to live and work with many challenges to be faced.

- Ionising radiations – the Earth's atmosphere and magnetic field protect us from dangerous ionising radiations such as energetic protons from the Sun and cosmic rays from outer space. The walls of a spacecraft are able to absorb the energetic protons but further research is required to find effective ways to protect astronauts from cosmic rays, which have much higher energies.
- Providing life support systems – energy is required to provide clean air, water, heating and cooling and to power electrical equipment on board the spacecraft. Solar cells can convert light (from the Sun) into the electrical energy required. However, as the spacecraft travels further from the Sun, less light will be received by these cells and a larger number (greater area) of solar cells would be required to provide a constant supply of electrical energy.
- A constant pressure and temperature within the spacecraft is required – spacecraft have to be able to withstand the pressure differential between atmospheric pressure inside the spacecraft and the near perfect vacuum outside and the extreme temperature variations, approximately 1000 °C, between regions in sunlight and shadow.
- Spacecraft have to be engineered to withstand collisions with tiny high-speed dust particles and space junk and the forces involved in lift-off from and re-entry through the Earth's atmosphere.

Figure 12.8 Chip in the window of the International Space Station probably caused by a small metal fragment of space junk

Benefits of space exploration

Space exploration has greatly increased our knowledge of the universe. However, the technology involving satellites and the use we make of them has probably had the greatest impact on our lives on Earth.

Satellites are used in the following areas:

- communications – 'instant' round-the-world audio and visual communication
- weather – allowing better weather prediction
- position – global positioning satellites (GPS) 'tell us' where we are
- environmental – vegetation monitoring, geological mapping and mineral prospecting, atmospheric chemistry, water vapour content, and ocean and land surface temperatures can all be monitored and measured using signals of different wavelengths coming from Earth

Figure 12.9 Scotland viewed from space

Figure 12.10 The Hubble telescope

Rockets and satellites

How do physicists ensure that the signals collected from space are of the widest range and highest quality?

Rockets are used to launch satellites into space. These satellites either orbit the Earth outside its atmosphere or are sent out into space.

Rocket development

In 1926, the American physicist Robert Goddard, who pioneered the use of liquid fuels to power rockets, launched the world's first liquid-propelled rocket. By 1944, during the Second World War, the Germans had produced a rocket, called the V2, which was able to fly from Holland to London, a distance of 160 km, in 6 minutes at an average speed of 900 m s^{-1} (2000 mph). After the war, larger and more powerful rockets were developed in order to put objects, mainly satellites and telescopes, into space.

Figure 12.11 Rockets at the NASA space museum

The principle of the rocket is simple: the rocket pushes hot gases downwards (action), the exhaust gases push the rocket upwards (reaction). This is Newton's third law.

The hot gases are created by burning a fuel – most rockets use liquid hydrogen. A large supply of oxygen is needed to burn the fuel and this is carried in the form of liquid oxygen. The fuel and oxygen mix and burn in a combustion chamber. The hot gases produced expand rapidly and are forced through the nozzle of the rocket. A diagram of a simplified rocket is shown in Figure 12.12.

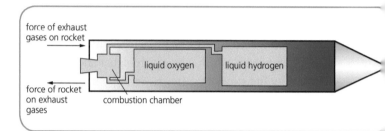
Figure 12.12 A simplified rocket

Lift-off

When a space rocket is launched, provided the force that the exhaust gas exerts on the rocket (thrust) is greater than the weight of the rocket then the rocket will accelerate upwards. These vertical forces acting on the rocket are shown in Figure 12.13.

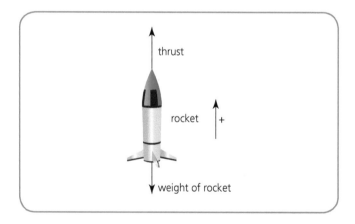
Figure 12.13 Vertical forces acting on a rocket at lift-off

$$\text{unbalanced force} = \text{thrust} + (- \text{weight of rocket})$$

$$\text{acceleration of rocket} = \frac{\text{unbalanced force}}{\text{mass of rocket}}$$

However, the rocket must be travelling at a speed greater than 11 000 m s^{-1} to escape the Earth's gravitational pull. To allow the rocket to reach this speed, the rocket's mass is reduced by ditching used or redundant parts, such as empty fuel tanks and the lower stage of the rocket. The rocket now exerts the same force (thrust) but on a smaller mass. This means that the acceleration of the remaining rocket increases.

In 'deep' outer space, where the gravitational field strength is zero, the weight of the spacecraft is zero. In this case:

unbalanced force on rocket = thrust

When manoeuvring in space, small thruster rockets are used instead of the main engine as only a small force is required. Manoeuvring is required when docking (joining together) with another spacecraft such as the International Space Station (ISS) in order to replenish supplies and exchange crew members and when preparing to re-enter Earth's atmosphere.

Weightlessness

In television programmes or films, when we see astronauts floating around in a spacecraft we say that the astronauts are 'weightless'. But is this true?

Worked example

Example

A rocket and spacecraft have a total mass at lift-off of 9.5×10^5 kg. The rocket exerts a thrust of 1.4×10^7 N at lift-off.
a) Calculate the acceleration of the rocket at lift-off.
b) The spacecraft journeys into deep space. During a manoeuvre to alter course, a manoeuvring rocket is switched on. This rocket has a thrust of 8.0 N. The spacecraft has a mass of 2500 kg. Calculate the acceleration of the spacecraft.

Solution

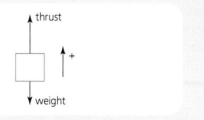

Figure 12.14

a) F_{un} = thrust + (− weight of rocket)
$F_{un} = 1.4 \times 10^7 - mg = 1.4 \times 10^7 - (9.5 \times 10^5 \times 9.8)$
$= 4.69 \times 10^6$ N

$F_{un} = ma$

$4.69 \times 10^6 = 9.5 \times 10^5 \times a$

$a = \dfrac{4.69 \times 10^6}{9.5 \times 10^5} = 4.9\,\mathrm{m\,s^{-2}}$

b) $F_{un} = ma$

$8 = 2500 \times a$

$a = \dfrac{8}{2500} = 3.2 \times 10^{-3}\,\mathrm{m\,s^{-2}}$

Weight is a force that pulls you towards the centre of the Earth – it pulls you down towards the ground. You exert a force on the ground and the ground exerts an equal but opposite force on you. You feel the effect of the ground supporting you and this is what makes you aware of your 'weight'.

Your weight is the pull of the Earth on your body ($W = mg$). It does not change unless either your mass or the gravitational field strength changes.

As a spacecraft rises from the surface of the Earth, the pull of the Earth's gravity on it gets smaller. Figure 12.15 shows how the gravitational field strength varies with height above the Earth.

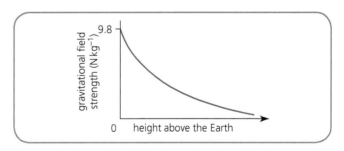

Figure 12.15 Graph showing how the gravitational field strength varies with height above the Earth

An astronaut in orbit (about 200 km above the surface of the Earth) still has weight since the gravitational field strength is about $7.0\,\mathrm{N\,kg^{-1}}$. This is strong enough to exert a force on the astronaut (their weight) and pull the astronaut towards the floor (towards the centre of the Earth) of the spacecraft. However, at the same time, there is a force on the spacecraft (its weight) pulling it towards the centre of the Earth. Both the spacecraft and the astronaut have the same acceleration towards the centre of the Earth. There is no contact between the astronaut and the floor of the spacecraft – the floor exerts no force on the astronaut – so the astronaut *appears* to be weightless.

The spacecraft and astronaut are two projectiles in a continual state of free fall – both are accelerating towards the Earth with the same acceleration. This is called 'weightlessness'.

True weightlessness can only be experienced in a region where the gravitational field strength is zero, i.e. in what is called 'deep' space.

Satellites in space

The period of a satellite is the time taken by the satellite to complete one orbit around the Earth. Table 12.1 shows the period of a satellite at different altitudes (height above sea level of the Earth).

Period of satellite (hours)	Altitude of orbit (km)
1.5	250
2.0	1 700
6	11 000
12	20 000
24	36 000

Table 12.1

As the altitude of a satellite increases, its period increases. To an observer on the Earth, as the altitude of a satellite increases then the slower the satellite appears to move. When a satellite is at an altitude of 36 000 km it completes one orbit in the same time as the Earth (24 hours) and therefore appears to remain stationary above a point on the Earth. This is called a **geostationary satellite**. Geostationary satellites are in orbit above the equator.

Hubble Space Telescope (HST)

This was launched in 1990 and orbits the Earth every 97 minutes. Electrical power is supplied by an array of photocells. Adjustments can be made by astronomers on the Earth to point the telescope in the desired direction. Since the Hubble telescope is in space, it detects signals from the faintest stars without the absorption or distortion caused by the Earth's atmosphere.

Images obtained by the telescope are helping physicists improve their estimates of the age of the universe. Observations of distant galaxies are providing evidence for the existence of black holes.

International Space Station (ISS)

Work on building the International Space Station started in 1998. This involved 16 countries. The crew based on the space station carry out long-term experiments on areas including materials, life sciences and medical research. It is hoped that the 'free fall' environment of the space station will allow scientific research to produce materials and drugs that have a greater purity compared with those produced on Earth.

Cosmic Background Explorer (COBE)

COBE was launched in 1989 and was a satellite dedicated to cosmology. Its job was to investigate the cosmic microwave background radiation in the universe. This work provided evidence to support the Big Bang theory of the universe.

Wilkinson Microwave Anisotropy Probe (WMAP)

WMAP was launched in 2001 and was used to measure small variations in the temperature of the Big Bang's remaining heat in the cosmic microwave background radiation. This succeeded the COBE space mission.

Global Positioning System (GPS)

This is a system of satellites orbiting the Earth. They continually transmit radio signals. A computer in a car or in a handset picks up these signals. The computer can calculate the time taken for the signal to reach it. By taking readings from a number of satellites, the computer can quickly calculate its own position on the Earth.

Search for extraterrestrial intelligence (SETI)

SETI is the name for a number of activities that people use to search for intelligent extraterrestrial life. This involves monitoring electromagnetic radiation for signs of transmissions from civilisations on other worlds.

Re-entry

When a spacecraft re-enters the Earth's atmosphere, it is travelling at high speed. As it collides with the air particles making up the atmosphere, a large resistive force acts on the spacecraft. The work done by this resistive force changes most of the kinetic energy of the spacecraft into heat. A large amount of this heat is 'lost' to the surroundings but some is absorbed by the spacecraft. Without special protection from this heat, the spacecraft would be destroyed.

Apollo

The Apollo space capsule used a heat shield made mainly from stainless steel. Most of the heat shield was vaporised (ablated) with the intense heat generated during re-entry, but this protected the capsule and the astronauts inside. After entering the atmosphere, parachutes were deployed to further slow the capsule down and allow it to make a gentle 'splashdown' in the sea.

Since the heat shield was largely 'destroyed' during re-entry and could not be replaced on the capsule, Apollo space capsules could not be re-used.

The Apollo missions allowed 12 men to walk on the surface of the Moon. The first Apollo spacecraft to land on the Moon, *Apollo 11*, landed in July 1969. The last mission was *Apollo 17* which took place in December 1972.

Space shuttle

The space shuttle used special tiles made from a silica compound to protect it from the intense heat of re-entry. The tiles transferred heat away quickly. The space shuttle was a reusable 'plane' that made an unpowered glide through the atmosphere and landed on a runway using its wheels just like an ordinary aircraft.

Key facts and physics equations: space exploration

- The period of a satellite depends on its altitude; the higher the orbit, the longer the period.
- Geostationary satellites have an altitude of 36 000 km and a period of 24 hours.

End-of-chapter questions

Information, if required, for use in the following questions can be found on the *Data Sheet* on page 170.

1 A rocket is ready for lift-off from the Earth.
   Draw a box to represent the rocket.
   Draw and name the vertical forces acting on the rocket at lift-off.

2 The mass of a rocket is 1500 kg. The rocket is to lift off from the Earth. The thrust exerted by the rocket at lift-off is 25 000 N.
   a) Calculate the weight of the rocket at lift-off.
   b) Calculate the acceleration of the rocket at lift-off.

3 The mass of a rocket is 2.2×10^6 kg. The rocket is launched vertically from the surface of the Earth. The rocket motors exert a constant thrust of 3.0×10^7 N.
   a) Calculate the unbalanced force on the rocket at lift-off
   b) Calculate the acceleration of the rocket at lift-off.
   c) A short time after lift-off, the acceleration of the rocket is greater than in b). Explain why the acceleration of the rocket increases as it rises.

4 A spacecraft blasts off from the surface of the Moon. The mass of the spacecraft is 4500 kg. The thrust exerted by the rocket of the spacecraft is 16 kN.
   a) Calculate the weight of the spacecraft on the Moon.
   b) Calculate the acceleration of the spacecraft on lift-off from the Moon.

5 A meteorite, made of iron, enters the upper atmosphere of the Earth. The meteorite is travelling at 15 000 m s^{-1}. The mass of the meteorite is 20 g.
   a) Calculate the kinetic energy of the meteorite when travelling at 15 000 m s^{-1}.
   b) If all the kinetic energy of the meteorite is converted into heat, calculate the rise in temperature of the meteorite.
   c) Explain why it is unlikely that the meteorite would reach the surface of the Earth.

6 During a tennis tournament in Chicago, microwave signals are sent from America to Britain by satellite.

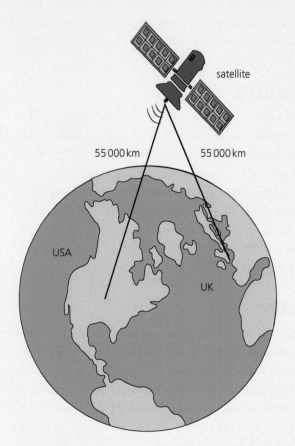

Figure 12.6

The distance from America to the satellite is 55 000 km. The microwave signal travels the same distance from the satellite to Britain. Calculate the time taken for the signal to be sent directly from America to Britain.

Cosmology

At the end of this chapter you should be able to:
1 Use the term light year correctly in context.
2 Convert between light years and metres.
3 State that there exists a family of waves with a wide range of wavelengths, which all travel at the speed of light.
4 Classify as members of the electromagnetic spectrum the following radiations: radio waves, television waves, microwaves, infrared, visible light, ultraviolet, X-rays, gamma rays.
5 List the radiations above in order of wavelength (and frequency).
6 Give an example of a detector for each of the radiations above.
7 State that telescopes can be designed to detect members of the electromagnetic spectrum.
8 State that different colours of light correspond to different wavelengths.
9 State that white light can be split into different colours using a prism.
10 List the following colours in order of wavelength: red, orange, yellow, green, blue, indigo, violet.
11 Explain why different kinds of telescope are used to detect signals from space.
12 State that the line spectrum produced by a source provides information about the atoms within the source.

Cosmology

Cosmology is a theory about the origin and the nature of the universe.

The universe is thought to have begun at a point in space with an explosion or **Big Bang**. Do not imagine that the Big Bang was like the explosion from a giant firework that you could have stood to one side and watched. The Big Bang represents the start of space and time. There is no position in space that you can point to and say 'this is where the Big Bang happened'. It happened everywhere. There was no *before* the Big Bang, since time began with the Big Bang. All the material from the Big Bang was thrown out, in all directions, and has been expanding ever since. Galaxies in the expanding universe are like 'chocolate chips in a cookie' – the chips get further and further apart as the dough expands.

Will the universe continue to expand for ever? Physicists have shown that there is not enough **visible matter** in the universe to produce sufficient gravitational attraction (pull) to stop the expansion. Many physicists believe that the universe contains lots of **invisible dark matter**. Evidence of the existence of dark matter comes from the gravitational effect of its mass on visible galaxies. Just enough of this dark matter would stop the universe expanding. Too much dark matter could not only stop the universe expanding but could put it into reverse – the universe would contract back on itself and eventually another Big Bang would start the cycle off once more.

The Big Bang is thought to have occurred about 15 000 million years (15 billion years) ago.

The light year

Everything connected with space is on a truly vast scale. For instance, the distance between the Earth and Proxima Centauri (our nearest star after the Sun) is 4.1×10^{16} m! This is a truly enormous distance but compared with most stars in our solar system, Proxima Centauri is a star that is really close to the Earth. To cope with such large distances requires a 'new unit' for distance, the **light year**.

The light year is the distance travelled, in metres, by light, in air or a vacuum, in 1 year.

Speed of light $= 3 \times 10^8\,\mathrm{m\,s^{-1}} = 300\,000\,000\,\mathrm{m\,s^{-1}}$
Distance travelled by light in 1 second $= 3 \times 10^8\,\mathrm{m}$
$\dfrac{\text{Distance travelled}}{\text{by light in 1 year}} = 3 \times 10^8 \times \dfrac{\text{number of seconds}}{\text{in 1 year}}$
1 year $= 365$ days $= (365 \times 24)$ hours
$\qquad\qquad\quad = (365 \times 24 \times 60)$ minutes
$\qquad\qquad\quad = (365 \times 24 \times 60 \times 60)\,\mathrm{s}$
$\dfrac{\text{Distance travelled}}{\text{by light in 1 year}} = 3 \times 10^8 \times (365 \times 24 \times 60 \times 60)$
One light year $= 9.5 \times 10^{15}\,\mathrm{m}$

The light year is required as a unit for measuring distance because the distances involved in space are so large.

Table 13.1 shows the time taken for light to travel to the Earth from distant parts of space, the distance travelled in that time and the number of light years it represents.

Source of light	Time taken for light to travel to the Earth	Distance travelled	Number of light years
Sun	8 minutes	$1.4 \times 10^{11}\,\mathrm{m}$	1.5×10^{-5}
Proxima Centauri – nearest star after the Sun	4.3 years	$4.1 \times 10^{16}\,\mathrm{m}$	4.3
Other side of our galaxy – the Milky Way	100 000 years	$9.5 \times 10^{20}\,\mathrm{m}$	100 000
Andromeda galaxy	2 500 000 years	$2.4 \times 10^{22}\,\mathrm{m}$	2 500 000

Table 13.1

The electromagnetic spectrum

In space, light is not the only transverse wave that travels at a speed of $3 \times 10^8\,\mathrm{m\,s^{-1}}$. All the waves of the electromagnetic spectrum also travel at $3 \times 10^8\,\mathrm{m\,s^{-1}}$ ($300\,000\,000\,\mathrm{m\,s^{-1}}$) (see Chapter 6: Waves).

The components of the electromagnetic spectrum and their detectors are shown in Table 13.2.

Radio waves, television waves and microwaves are detected electronically and an aerial is required to convert these radiations into electric currents. Infrared is detected using a photodiode (the photodiode converts the radiation into an electric signal). Visible light is detected by the retina of the human eye. Ultraviolet can be detected by fluorescent materials (these are materials that glow when exposed to ultraviolet). Ultraviolet, X-rays and gamma rays can be detected using photographic film (which turns black or 'fogs' when exposed to the radiation); gamma rays can also be detected using a Geiger–Müller tube and counter.

	Wave	Detector
Long wavelength, low frequency	Radio	Aerial and radio receiver
	Television	Aerial and television receiver
	Microwaves	Aerial and microwave receiver
	Infrared	Photodiode and meter
	Visible light	Retina of eye, Photodiode and meter
	Ultraviolet	Fluorescent materials, Photographic film
	X-rays	Photographic film, X-ray intensifier
Short wavelength, high frequency	Gamma rays	Photographic film, Geiger–Müller tube and counter

Table 13.2

Detecting signals from space

Astronomers, in addition to observing the visible spectrum (using the eye), also observe radiations that range from long wavelength radio waves to short wavelength X-rays and gamma rays. Looking at objects in different parts of the electromagnetic spectrum reveals much more about them. Each type of radiation produces its own problems and to observe much of the electromagnetic spectrum requires special telescopes to be sent into space.

Radio telescopes

Large curved parabolic dishes (see Figure 13.1) made from unpolished metal, often mesh wire, collect and direct weak radio waves from a particular part of outer space onto an aerial. The aerial, at the focus of the dish, converts the radio waves into an electrical signal that is amplified before being analysed. Radio waves have long wavelengths. This means that the collecting surface need not be accurately shaped.

Figure 13.1 The 76 m Lovell telescope at the Jodrell Bank Observatory, UK

To see fine detail, the aperture (the dish diameter, the part that collects the waves) of a radio telescope should be as large as possible. However, single-dish radio telescopes cannot be made physically large enough for the detail required by scientists. The problem of size is overcome by using several smaller dishes in a line (see Figure 13.2). This is called **aperture synthesis**. The results are computerised, and apertures of many kilometres can be simulated.

Figure 13.2 The Very Large Array telescope, New Mexico

Microwaves

Astronomers can detect radiation from space that has a wavelength of a few millimetres, called **microwave radiation**. At ground level almost all microwave wavelengths from outer space have been absorbed by the Earth's atmosphere. As a result, microwave telescopes are sited on high mountains. As the wavelengths of these waves are so short, the dishes of millimetre-wave telescopes have to be accurately shaped. They have highly polished dishes and are housed in temperature-controlled buildings. This is because if the temperature changes, the shape of the dish will change and distort the image.

Infrared radiation

Infrared radiation arrives at the Earth from objects in space and provides astronomers with another source of information. Infrared radiation has a longer wavelength (lower frequency) than visible light and cannot be detected by the retina of the human eye.

Most infrared radiation is absorbed by the Earth's atmosphere, although there are several 'windows' where infrared radiation of a certain wavelength can penetrate through to high altitudes. As a result, infrared telescopes are sited on high mountains or on satellites above the Earth's atmosphere.

Infrared telescopes are able to 'see' stars that are hidden at visible wavelengths behind dusty regions of space. When the infrared radiation is detected by the telescope, the invisible radiation (to our eyes) is converted into a signal that can be viewed by humans.

Optical refracting telescope (visible light telescope)

Up until the invention of the light telescope in 1609, astronomers could only look at the stars with the naked eye. The light telescope allowed astronomers to see further, and in greater detail, into space.

A light telescope (Figure 13.3) consists of two accurately made glass lenses enclosed in a light-tight tube. Faint light waves from the star or object being viewed pass through a lens, called the **objective lens**. Another lens, called the **eyepiece lens**, is used to focus the light from the objective lens onto your eye. The two lenses, when focused, make distant objects look larger and nearer.

The brightness of the image obtained from a light telescope is determined by the diameter of the objective lens (also known as the **aperture**). Since light from a distant star is so faint, large-diameter objective lenses are desirable to give a bright, detailed image. However, as light waves have very short wavelengths, the surfaces of glass telescope lenses must be very accurately shaped. This means that large-diameter lenses are difficult and expensive to produce.

Large optical telescopes are located on high mountains away from light and atmospheric pollution.

Ultraviolet

Some wavelengths of ultraviolet radiation from outer space can be detected at ground level but most ultraviolet radiation is absorbed by the atmosphere. As a result, ultraviolet observations are made by satellites above the Earth's atmosphere.

Major sources of ultraviolet radiation include hot, massive stars and the cores of active galaxies.

Ultraviolet radiation is invisible to the human eye. When the ultraviolet radiation is detected, the invisible radiation (to our eyes) is converted into a signal that can be viewed by humans.

X-rays and gamma rays

Because the Earth's atmosphere absorbs X-rays and gamma rays from outer space, it is not possible to build ground-based telescopes for these wavelengths. X-ray and gamma ray observations are made by satellites above the Earth's atmosphere.

Most ordinary stars emit only weak X-rays. Strong X-ray sources are observed from extremely hot gas, active galaxies and rich clusters of galaxies.

Major sources of cosmic gamma rays include solar flares, pulsars (rapidly rotating neutron stars detected by the pulses of radiation they give off), remnants of supernovae (stars that explode, throwing out most of their material and brightening by a factor of about a million) and active galaxies.

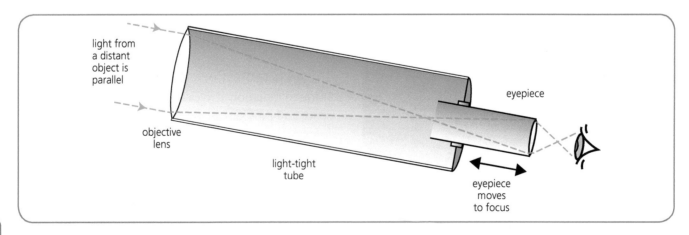

light from a distant object is parallel

objective lens

light-tight tube

eyepiece

eyepiece moves to focus

Figure 13.3 A (visible light) telescope

Temperature of stars
Colour and temperature

Consider a filament lamp connected to a power supply. When the power supply is switched on and the voltage is slowly increased, the current in the filament increases. When the filament becomes hot enough it emits a dull red glow – it is red hot. As the temperature of the filament rises further, not only does it emit more light but the colour of the light it emits changes – becoming orange then yellow. Eventually, when the filament is hot enough, it emits white light (the filament is white hot) or even light that is distinctly blue in colour. This applies to all hot substances, including stars.

As a result, astronomers can judge the temperature of the surface of a star by observing the colour of light emitted. Table 13.3 shows the colour of light emitted by four stars.

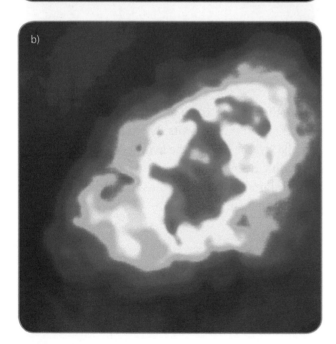

Figure 13.4 a) Composite image of the Crab Nebula taken by infrared (red area), visible (blue and green) and X ray (central light blue area) telescopes; **b)** A radio image of the Crab Nebula

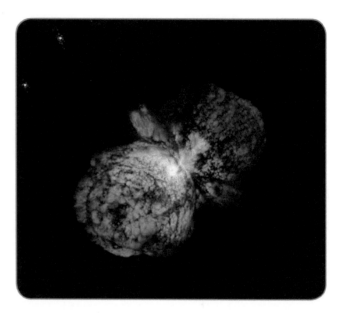

Figure 13.5 Image of Eta-Carinae taken by the Hubble telescope

Name of star	Betelgeuse	Arcturus	Sirius	Rigel
Colour of light emitted	Red	Orange–red	White	Bluish white
Relative temperature	Coolest			Hottest

Table 13.3

A more scientific way to judge the temperature of any hot object is to analyse the light emitted using a spectrometer or spectroscope. A spectrometer uses a prism to split up the light from the hot object (or star) into a **spectrum** of different wavelengths (colours).

Continuous spectrum

When a beam of white light (from a hot filament lamp) passes through a prism, the different wavelengths (colours) of light slow down by different amounts causing the beam to fan out into the different colours (Figure 13.6). The light is bent again on leaving the prism so producing a spectrum of colours – like a rainbow. The continuous spectrum of white light consists of: red, orange, yellow, green, blue, indigo and violet.

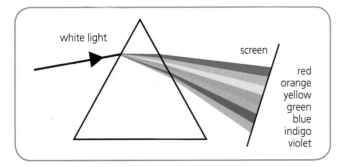

Figure 13.6 A continuous spectrum

Red light has the longest wavelength and the lowest frequency.

Violet light has the shortest wavelength and the highest frequency.

Line spectra

Each element of the periodic table emits its own set of unique wavelengths of visible light. These wavelengths are called **emission lines** and produce what is called a **line spectrum**. They are the 'finger print' for that element. The line spectra for a number of elements are shown in Figure 13.7.

This means that astronomers can identify the elements present in a distant star by viewing its line spectra.

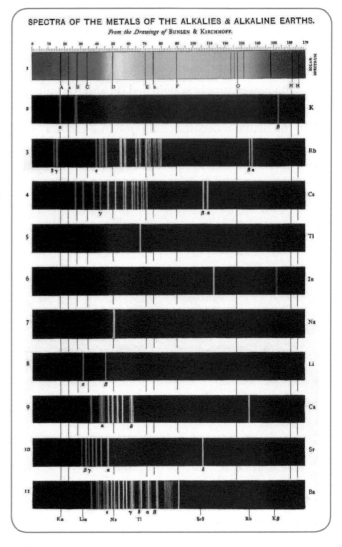

Figure 13.7 Line spectra for different elements

Compare the line spectrum of the unknown source with the spectra from the elements shown in Figure 13.8. Notice that *all* the lines in the hydrogen and helium spectra coincide in the unknown source. This means that the unknown source contains the elements hydrogen and helium.

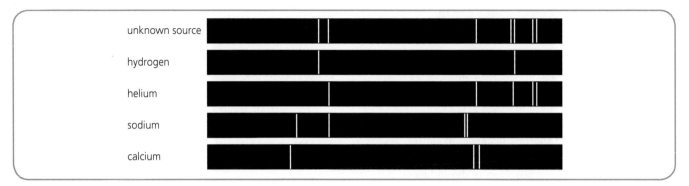

Figure 13.8 Spectra for an unknown source and the spectra from known sources

Key facts and physics equations: cosmology

- A light year is the distance travelled by light in 1 year.
- All members of the electromagnetic spectrum are transverse waves and are transmitted through a vacuum or air at a speed of $3 \times 10^8 \, \text{m s}^{-1}$.
- The members of the electromagnetic spectrum: radio waves (longest wavelength, lowest frequency), television waves, microwaves, infrared radiation, visible light, ultraviolet radiation, X-rays, gamma rays (shortest wavelength, highest frequency).
- Different members of the electromagnetic spectrum require different detectors.
- Different wavelengths of light correspond to different colours.
- The visible spectrum consists of the colours: red, orange, yellow, green, blue, indigo, violet.
- A line spectrum produced by a source can be used to identify the elements making up the source.

End-of-chapter questions

Information, if required, for use in the following questions can be found on the *Data Sheet* on page 170.

1 Sirius is a star that is 8.6 light years away from Earth.
   a) What is meant by the term *light year*?
   b) Calculate the shortest distance, in metres, between Sirius and the Earth.

2 The electromagnetic spectrum is shown below.

Radio	Television	P	Infrared	Visible	Q	X-rays	Gamma rays

   a) Name radiations P and Q.
   b) Which member of the electromagnetic spectrum has the shortest wavelength?

3 An astronomer observes the colours of light emitted by four stars:
   • Star W emits orange–red
   • Star X emits bluish white
   • Star Y emits yellow
   • Star Z emits red.
   Place the stars in order of their temperature, starting with the coolest star.

4 Light from a distant star is viewed through a telescope and analysed with a spectrometer. A set of coloured lines on a dark background is seen. What information about this star can be obtained from these lines?

5 State a detector used for each of the following radiations:
   a) X-rays
   b) ultraviolet
   c) microwaves.

Exam practice for Chapters 9–13

Information, if required, for use in the following questions can be found on the *Data Sheet* on page 170.

1 a) Lynne cycles to and from school. The journey takes her along a short stretch of straight road, past her friend Susan's house. Susan decides to measure Lynne's average speed along the straight road. Susan uses a stopwatch and a measuring tape. Describe how Susan could measure Lynne's average speed as she travels along the straight road.

  b) During a part of her journey, Lynne has to carry her bicycle up a flight of stairs. There are ten steps. Each step is 80 mm high. The mass of her bicycle is 15 kg.
    i) What is the minimum force required to lift the bicycle?
    ii) Calculate the minimum work done in raising the bicycle up the flight of ten steps.

2 A train is travelling along a straight, level track at constant speed. The driver sees that the signal ahead is at danger. He applies the train's brakes. The graph in Figure E3.1 shows how the speed of the train varies with time from the instant the signal is seen by the driver.

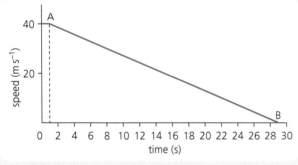

Figure E3.1

  a) Calculate the acceleration of the train between A and B.
  b) When the driver first sees the signal, the train is 620 m away from it. Does the train stop in this distance? You **must** clearly show the working that leads to your answer.

3 a) The mass of a ship is 240 000 kg. The ship is towed in a straight line at a constant speed of 1.5 m s⁻¹ by a tug. When pulling the ship, the tug exerts a constant force of 1200 N on the towing cable. What is the size of the resistive force of the water opposing the motion of the ship?

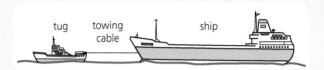

Figure E3.2

  b) The towing cable is now released. The ship continues to travel in a straight line. It slows down and comes to rest. The resistive force of the water on the ship remains constant as the ship comes to rest.
    i)   Calculate the acceleration of the ship.
    ii) After the towing cable is released, how long does it take for the ship to come to rest?
    iii) After the towing cable is released, how far does the ship travel?

4 A platform is 0.5 m above the ground. An elevator is used to raise, at constant speed, bales of straw from the platform on to a stack. The stack is 5.5 m above the ground. The mass of each bale is 25 kg.

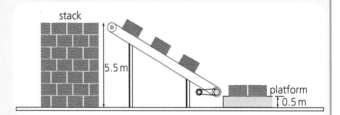

Figure E3.3

  a) Calculate the gain in gravitational potential energy of a bale when it is lifted from the platform to the stack.
  b) The elevator raises 20 bales of straw to the stack in 5 minutes.
    i) Calculate the minimum power of the electric motor of the elevator during this time.
    ii) Give **one** reason why the power of the electric motor is greater than the value calculated in b) i).

5 Figure E3.4 shows a pendulum bob at point X, its rest position. The mass of the bob is 0.2 kg. The pendulum bob is now pulled to point Y. Point Y is 0.3 m vertically above the rest position.
  a) Calculate the gain in gravitational potential energy of the bob when it is moved from X to Y.
  b) The pendulum bob is released from Y and swings to and fro until it comes to rest.

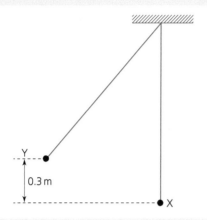

Figure E3.4

    i) Describe the energy changes that take place as the pendulum bob swings from Y to X.

    ii) Show that the maximum possible speed of the pendulum bob is 2.42 m s⁻¹. State any assumption you make in your calculation.

6 A girl starts from rest at point P on a ski slope. She accelerates uniformly down the slope to point Q. She then slows down as she moves along the level section and stops at point R. Point P is 6.0 m vertically above point Q. The mass of the girl is 45 kg.

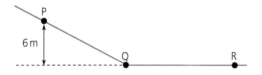

Figure E3.5

  a) Calculate the change in gravitational potential energy of the girl as she moves from P to Q.

  b) Assuming that all her gravitational potential energy is transferred to kinetic energy, calculate the speed of the girl at Q.

  c) As the girl travels from Q to R, there is a constant frictional force of 15 N acting on her. Calculate the distance travelled by the girl as she travels along the level section between Q and R.

7 During a sailing competition, a boy competes in a race. The graph in Figure E3.6 shows how the velocity of the boat varies with time during part of the race.

  a) i) Calculate the acceleration of the boat during the first 30 seconds of the race.

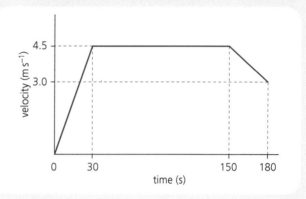

Figure E3.6

    ii) Calculate the distance travelled by the boat during the race.

    iii) Calculate the average speed of the boat during the race.

  b) Figure 3.7 shows the horizontal forces acting on the boat during the race.

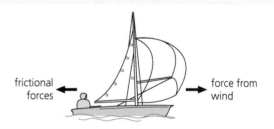

Figure E3.7

    i) During the first 30 seconds of the race are these forces balanced or unbalanced? Justify your answer.

    ii) During the race, between 30 s and 150 s, are these forces balanced or unbalanced? Justify your answer.

8 A ball is dropped vertically from a window of a house onto the ground below. The graph in Figure E3.8 shows how the velocity of the ball varies with time for the first 0.8 s of its motion.

  a) Describe the motion of the ball during:

    i) OA

    ii) BC.

  b) Calculate the height above the ground from which the ball is dropped.

  c) Calculate the height to which the ball rebounds.

  d) Why is your answer to c) smaller than the answer to b)?

  e) At what time does the ball hit the ground? Justify your answer.

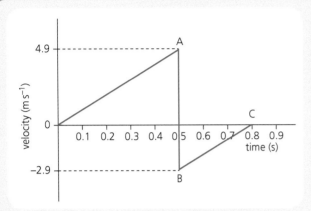

Figure E3.8

The ball hits the ground 0.70 s later.
The graphs in Figure E3.10 show how the horizontal velocity and the vertical velocity of the ball vary with time during the 0.70 s.

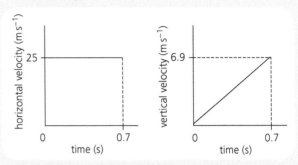

Figure E3.10

9 A boat is travelling south at a constant speed of 12 m s⁻¹. The tide is moving from east to west at a constant speed of 6.0 m s⁻¹. The mass of the boat is 14 000 kg.
  a) Calculate the resultant velocity of the boat.
  b) The boat enters an area where there is no tide. The boat travels at a constant speed of 12 m s⁻¹ in a straight line. The frictional force exerted by the water is 1200 N. What is the forward force acting on the boat? Explain your answer.

10 A ball rolls along a flat, horizontal roof top at a constant speed of 2.5 m s⁻¹. The ball falls off the end of the roof top and hits the ground 3.2 s later. The mass of the ball is 0.3 kg. The effects of air resistance on the ball can be ignored.
  a) In the time taken to fall to the ground, calculate the horizontal distance travelled by the ball.
  b) Calculate the vertical speed of the ball just before it hits the ground.
  c) Calculate the height that the ball falls through.

11 When serving a tennis ball, a tennis player hits the ball horizontally as shown in Figure E3.9.

Figure E3.9

The effects of air resistance on the ball can be ignored.
  a) Explain why the shape of the path taken by the ball is curved.
  b) Calculate the horizontal distance travelled by the ball.
  c) Calculate the height the ball is hit from.

12 A polar orbiting satellite takes a shorter time to orbit the Earth than a communications satellite.
  a) How do the altitudes of these satellites compare?
  b) A communications satellite is in a geostationary orbit.
   i) State what is meant by the term *geostationary orbit*.
   ii) The communications satellite is 36 000 km above the surface of the Earth. A radio signal is sent from the ground to the satellite and then relayed back to the ground.
      Calculate the minimum time taken for the radio signal to make this journey.

13 A satellite is to be launched from the Earth in order to investigate electromagnetic radiation in space. The satellite is carried into orbit using a rocket. The following is information about the rocket and satellite:
  • Mass of rocket = 2.1 × 10⁶ kg
  • Mass of satellite = 2.3 × 10³ kg
  • Thrust exerted by rocket engines at lift-off = 2.5 × 10⁷ N
  a) Explain how the rocket is able to lift-off from the Earth.
  b) Calculate the acceleration of the rocket and satellite at lift-off.
  c) At a certain time during the launch, the speed of the rocket and satellite is 1200 m s⁻¹. Calculate the kinetic energy of the **satellite** at this time.

d) When placed in orbit, the satellite is 900 km above the surface of the Earth. To test a piece of equipment a microwave signal of wavelength 30 mm is sent to the satellite. Calculate the time taken for the signal to travel the 900 km to the satellite.

e) One of the probes on the satellite detects an infrared signal with a frequency of 1.06×10^{15} Hz. Calculate the wavelength of the signal detected.

f) To power the equipment on the satellite, an array of solar cells is used. Under certain conditions, one of the solar cells produces a voltage of 0.8 V. The current from the cells is 60 mA. Calculate the power produced by the solar cell.

14 A space probe is investigating the atmosphere of a planet. The mass of the probe is 750 kg. The speed of the probe is 380 m s^{-1}. As the probe moves through the atmosphere its speed is reduced to 170 m s^{-1} due to the resistive force from the atmosphere.

a) Calculate the change in kinetic energy of the probe.

b) The specific heat capacity of the material of the space probe is 1400 J kg^{-1} °C^{-1}. Calculate the maximum rise in temperature of the space probe.

c) During the construction of the space probe, a sample of the probe material is heated to its melting point. The mass of the sample is 600 g. The power rating of the heater used to supply energy to the sample is 2400 W. At its melting point, the time taken to melt the sample is 130 s.

i) Calculate the specific latent heat of fusion of the sample.

ii) Explain whether your answer for the specific latent heat of fusion is too high or too low.

Useful Physics equations

$E_P = mgh$

$E_K = \dfrac{1}{2}mv^2$

$Q = It$

$V = IR$

$R_T = R_1 + R_2 + \ldots$

$\dfrac{1}{R_T} = \dfrac{1}{R_1} + \dfrac{1}{R_2} + \ldots$

$V_2 = \left(\dfrac{R_2}{R_1 + R_2}\right)V_s$

$\dfrac{V_1}{V_2} = \dfrac{R_1}{R_2}$

$P = \dfrac{E}{t}$

$P = IV$

$P = I^2R$

$P = \dfrac{V^2}{R}$

$E_h = cm\Delta T$

$p = \dfrac{F}{A}$

$\dfrac{pV}{T} = \text{constant}$

$p_1V_1 = p_2V_2$

$\dfrac{p_1}{T_1} = \dfrac{p_2}{T_2}$

$\dfrac{V_1}{T_1} = \dfrac{V_2}{T_2}$

$d = vt$

$v = f\lambda$

$T = \dfrac{1}{f}$

$A = \dfrac{N}{t}$

$D = \dfrac{E}{m}$

$H = Dw_R$

$\dot{H} = \dfrac{H}{t}$

$s = vt$

$d = \bar{v}t$

$s = \bar{v}t$

$a = \dfrac{v - u}{t}$

$W = mg$

$F = ma$

$E_W = Fd$

$E_h = ml$

Data sheet

Speed of light in materials

Material	Speed in $m\,s^{-1}$
air	3.0×10^8
carbon dioxide	3.0×10^8
diamond	1.2×10^8
glass	2.0×10^8
glycerol	2.1×10^8
water	2.3×10^8

Gravitational field strengths

	Gravitational field strength on the surface in $N\,kg^{-1}$
Earth	9.8
Jupiter	23
Mars	3.7
Mercury	3.7
Moon	1.6
Neptune	11
Saturn	9.0
Sun	270
Uranus	8.7
Venus	8.9

Specific latent heat of fusion of materials

Material	Specific latent heat of fusion in $J\,kg^{-1}$
alcohol	0.99×10^5
aluminium	3.95×10^5
carbon dioxide	1.80×10^5
copper	2.05×10^5
iron	2.67×10^5
lead	0.25×10^5
water	3.34×10^5

Specific latent heat of vaporisation of materials

Material	Specific latent heat of vaporisation in $J\,kg^{-1}$
alcohol	11.2×10^5
carbon dioxide	3.77×10^5
glycerol	8.30×10^5
turpentine	2.90×10^5
water	22.6×10^5

Speed of sound in materials

Material	Speed in $m\,s^{-1}$
aluminium	5200
air	340
bone	4100
carbon dioxide	270
glycerol	1900
muscle	1600
steel	5200
tissue	1500
water	1500

Specific heat capacity of materials

Material	Specific heat capacity in $J\,kg^{-1}\,°C^{-1}$
alcohol	2350
aluminium	902
copper	386
glass	500
ice	2100
iron	480
lead	128
oil	2130
water	4180

Melting and boiling points of materials

Material	Melting point in °C	Boiling point in °C
alcohol	−98	65
aluminium	660	2470
copper	1077	2567
glycerol	18	290
lead	328	1737
iron	1537	2737

Radiation weighting factors

Type of radiation	Radiation weighting factor
alpha	20
beta	1
fast neutrons	10
gamma	1
slow neutrons	3
X-rays	1

SI prefixes and multiplication factors

Prefix	Symbol	Factor
giga	G	$10^9 = 1\,000\,000\,000$
mega	M	$10^6 = 1\,000\,000$
kilo	k	$10^3 = 1000$
milli	m	$10^{-3} = 0.001$
micro	μ	$10^{-6} = 0.000\,001$
nano	n	$10^{-9} = 0.000\,000\,001$

Direct and indirect proportion

Direct proportion (or direct variation)

Imagine that you are doing an experiment to establish a relationship between two quantities. You record the measurements taken in a table as shown below.

R (tore)	S (lumens)	R/S
0	0	-
5	2	2.5
10	4	2.5
15	6	2.5
20	8	2.5
25	10	2.5
30	12	2.5

Since R goes up as S goes up, see what $\dfrac{R}{S}$ gives.

Notice that $\dfrac{R}{S}$ always gives the same answer, i.e. it gives a constant.

Notice also that if S is doubled, R is doubled and if S is tripled, R is tripled and so on.

The factor changed by the person doing the experiment – the independent variable – is plotted on the x-axis. The other variable – the dependent variable – is plotted on the y-axis.

The graph of R against S is a straight line through the origin.

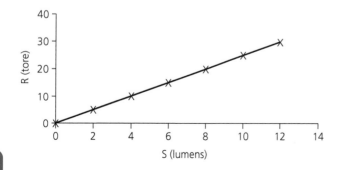

This means that R is directly proportional to S (if you double R you double S and so on).

- R is directly proportional to S
- $R \propto S$
- $R = kS$ ($\propto$ can always be replaced by $= k$ where k is a constant)
- $\dfrac{R}{S} = k$

In this case, $\dfrac{R}{S} = k = 2.5$

Indirect (or inverse) proportion

Imagine that you are doing an experiment to establish the relationship between two quantities P and Q. You record the measurements taken in a table as shown below.

P (poundels)	Q (tesla)	P × Q
64	3	192
32	6	192
16	12	192
8	24	192
4	48	192
2	96	192

Since P goes down as Q goes up, see what $P \times Q$ gives.

Notice that $P \times Q$ always gives the same answer, i.e. gives a constant.

Notice also that if Q doubles, P halves and if Q goes up by a factor of 4, then P quarters.

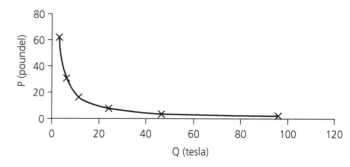

The graph is a smooth curve which does not cut either axis. This suggests that P is inversely proportional to Q (i.e. double Q, you half P). A graph of P against $\dfrac{1}{Q}$ should now be drawn.

P (poundels)	Q (tesla)	1/Q (1/tesla)
64	3	0.33
32	6	0.17
16	12	0.08
8	24	0.04
4	48	0.02
2	96	0.01

The graph of P against $\dfrac{1}{Q}$ is a straight line through the origin.

This means that :

- P is directly proportional to $\dfrac{1}{Q}$
- P is inversely proportional to Q
- $P \propto \dfrac{1}{Q}$ ($\propto$ means is directly proportional to)
- $P = k \times \dfrac{1}{Q}$
- $P \times Q = k$ ($=$ constant)

In this case $P \times Q = k = 192$

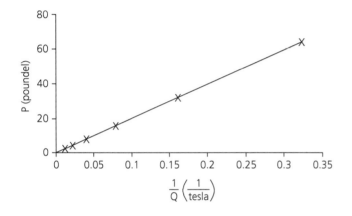

Index

Answers

Answers to Section 1

1 Electrical circuits (pages 15–16)

1 Note: t must be in seconds; 3 minutes $= (3 \times 60)\,$s
$Q = It = 5 \times (3 \times 60) = 900\,$C

2 a) Current is the charge transferred in 1 second.
   b) $Q = It$
   $800 = 2 \times t$
   $t = \dfrac{800}{2} = 400\,$s

3 Note: t must be in seconds;
   5 minutes $= (5 \times 60)\,$s
   $Q = It$
   $750 = I \times (5 \times 60)$
   $I = \dfrac{750}{300} = 2.5\,$A

4 9.0 V means that 9 joules of energy are given to each coulomb of charge as it passes through the supply.

5 See Figure 1.7 on page 6.

6 a) $A_1 = 0.05\,$A since current is the same at all points in a series circuit.
   $A_2 = 0.30\,$A as circuit current = sum of currents in the branches of a parallel circuit.
   $A_3 = 0.15\,$A as circuit current = sum of currents in the branches of a parallel circuit.
   b) $V_1 = 2.0\,$V as supply voltage = sum of voltages in a series circuit.
   $V_2 = 6.0\,$V as voltages are the same across components connected in parallel.
   $V_3 = 9.0\,$V as supply voltage = sum of voltages in a series circuit.

7 $R_T = R_1 + R_2 + R_3 = 47 + 100 + 150 = 297\,\Omega$

8 $\dfrac{1}{R_T} = \dfrac{1}{R_1} + \dfrac{1}{R_2} + \dfrac{1}{R_3} = \dfrac{1}{20} + \dfrac{1}{20} + \dfrac{1}{10}$

   $\dfrac{1}{R_T} = 0.050 + 0.050 + 0.100 = 0.200$

   $R_T = \dfrac{1}{0.200} = 5.0\,\Omega$

   As a check, R_T must be smaller than the smallest resistor, i.e. $10\,\Omega$.

9 For the resistors connected in parallel:

   $\dfrac{1}{R_{parallel}} = \dfrac{1}{R_1} + \dfrac{1}{R_2} = \dfrac{1}{60} + \dfrac{1}{30} = 0.017 + 0.033 = 0.050$

   $\dfrac{1}{R_{parallel}} = 0.050$

   $R_{parallel} = \dfrac{1}{0.050} = 20\,\Omega$

   As a check, $R_{parallel}$ must be smaller than $30\,\Omega$.
   $R_{XY} = 30 + R_{parallel} + 20 = 30 + 20 + 20 = 70\,\Omega$

10 See Figures 1.8 and 1.9 on pages 6 and 7.

11 Note: $1\,\text{k}\Omega = 1 \times 10^3\,\Omega = 1000\,\Omega$
   $V = IR = 0.012 \times 1000 = 12\,$V

12 $V = IR$
   $230 = 4.5 \times R$
   $R = \dfrac{230}{4.5} = 51.1\,\Omega$

13 $V = IR$
   $4.5 = I \times 18$
   $I = \dfrac{4.5}{18} = 0.25\,$A

14 Note: $1\,\text{k}\Omega = 1 \times 10^3\,\Omega = 1000\,\Omega$
   $R_T = R_1 + R_2 = 1000 + 3000 = 4000\,\Omega$

   $I_{circuit} = \dfrac{V_{supply}}{R_T} = \dfrac{9}{4000} = 0.00225\,$A

   $V_{3\,k\Omega} = (IR)_{3\,k\Omega} = 0.00225 \times 3000 = 6.75\,$V

2 Electrical energy and power (page 21)

1 a) $P = IV = 5.5 \times 230 = 1265\,$W
   b) $1265\,$J (as $P = \dfrac{E}{t}$ and $t = 1\,$s)

2 a) Electrical to heat and light
   b) $V = IR$
   $12 = I \times 3$
   $I = \dfrac{12}{3} = 4.0\,$A
   c) $P = IV = 4 \times 12 = 48\,$W
   (or $P = I^2R = 48\,$W or $P = \dfrac{V^2}{R} = 48\,$W)

3 $P = \dfrac{V^2}{R} = \dfrac{230^2}{17.6} = 3006\,\text{W}$

4 $P = \dfrac{V^2}{R}$

$50 = \dfrac{12^2}{R}$

$R = \dfrac{12^2}{50} = 2.9\,\Omega$

5 a) A power of 138 W means that 138 joules of energy is transferred each second.

   b) $P = I^2R$

   $138 = 0.6^2 \times R$

   $R = \dfrac{138}{0.36} = 383\,\Omega$

6 a) See Figure 2.1.

   b) Safety device, i.e. to prevent wiring from overheating.

7 a) 3 A

   b) 13 A

   c) 13 A

8 X = fuse, Y = lamp, Z = resistor

9 a) Electrons move in only one direction round the circuit.

   b) Electrons move in one direction, then in the opposite direction and then back again, i.e. electrons move to and fro.

3 Electrical components and electronic circuits (pages 34–36)

1 a) motor

   b) variable resistor

   c) microphone

   d) diode

   e) loudspeaker

   f) LDR

   g) MOSFET

   h) thermistor

   i) LED

   j) NPN transistor

   k) solar cell

   l) capacitor

   m) relay

2 a) sound to electrical

   b) light to electrical

   c) electrical to sound

   d) electrical to kinetic

   e) electrical to heat and light

   f) electrical to light

3 a) Note: 12 mA = 12×10^{-3} A = 0.012 A

   $V = IR$

   $6 = 12 \times 10^{-3} \times R$

   $R = \dfrac{6}{12 \times 10^{-3}} = 500\,\Omega$

   b) As temperature increases, resistance of thermistor decreases and therefore current increases – any reasonable value greater than 12 mA, e.g. 13 mA.

4 a)

   b) Since components are connected in series:

   $V_{\text{supply}} = V_{\text{LED}} + V_{\text{resistor}}$

   $6 = 1.75 + V_{\text{resistor}}$

   $V_{\text{resistor}} = 4.25\,\text{V}$

   Note: 11 mA = 11×10^{-3} A = 0.011 A

   $\dfrac{\text{voltage across}}{\text{resistor}} = \dfrac{\text{current in}}{\text{resistor}} \times \dfrac{\text{resistance of}}{\text{resistor}}$

   $V_{\text{resistor}} = (IR)_{\text{resistor}}$

   $4.25 = 11 \times 10^{-3} \times R$

   $R = \dfrac{4.25}{11 \times 10^{-3}} = 386\,\Omega$

5 a) $R_{\text{T}} = R_1 + R_2 = 6 + 30 = 36\,\Omega$

   $V_{\text{supply}} = I_{\text{circuit}} \times R_{\text{T}}$

   $6 = I_{\text{circuit}} \times 36$

   $I_{\text{circuit}} = \dfrac{6}{36} = 0.167\,\text{A}$

   $V_1 = I_{\text{circuit}} \times R_1 = 0.167 \times 6 = 1.0\,\text{V}$

   $V_2 = I_{\text{circuit}} \times R_2 = 0.167 \times 30 = 5.0\,\text{V}$

   b) Note: $4.0\,\text{k}\Omega = 4 \times 10^3\,\Omega = 4000\,\Omega$

   $2.0\,\text{k}\Omega = 2 \times 10^3\,\Omega = 2000\,\Omega$

   $R_{\text{T}} = R_1 + R_2 = 4000 + 2000 = 6000\,\Omega$

   $V_{\text{supply}} = I_{\text{circuit}} \times R_{\text{T}}$

   $12 = I_{\text{circuit}} \times 6000$

   $I_{\text{circuit}} = \dfrac{12}{6000} = 0.002\,\text{A}$

   $V_1 = I_{\text{circuit}} \times R_1 = 0.002 \times 4000 = 8.0\,\text{V}$

   $V_2 = I_{\text{circuit}} \times R_2 = 0.002 \times 2000 = 4.0\,\text{V}$

c) $R_T = R_1 + R_2 = 500 + 2000 = 2500\,\Omega$

$V_{supply} = I_{circuit} \times R_T$

$10 = I_{circuit} \times 2500$

$I_{circuit} = \dfrac{10}{2500} = 0.004\,A$

$V_1 = I_{circuit} \times R_1 = 0.004 \times 500 = 2.0\,V$

$V_2 = I_{circuit} \times R_2 = 0.004 \times 2000 = 8.0\,V$

6 As the light intensity increases, the resistance of the LDR decreases. Voltage across the LDR will decrease and so the voltage across the resistor will increase and so the voltmeter reading increases.

7 a) Increases
   b) Increase value of resistor *or* increase value of capacitor *or* increase value of supply voltage.

8 a) W = thermistor, X = resistor, Y = NPN transistor, Z = lamp
   b) An electronic switch

9 As the resistance of the thermistor decreases the voltage across the thermistor decreases. So the voltage across the resistor increases and when it is equal to or greater than a certain value, the transistor switches on and the lamp lights.

10 As the light level on the LDR increases, the resistance of the LDR decreases and the voltage across the LDR decreases. Therefore the voltage across the resistor increases and when it is equal to or greater than a certain value, the transistor switches on and the LED lights.

11

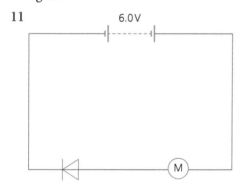

6.0V

M

12 a) $R_T = R_1 + R_2 = 10 + 90 = 100\,k\Omega = 100 \times 10^3\,\Omega$

$V_{supply} = I_{circuit} \times R_T$

$5 = I_{circuit} \times 100 \times 10^3$

$I_{circuit} = \dfrac{5}{100 \times 10^3} = 5 \times 10^{-5}\,A$

$V_{LDR} = I_{circuit} \times R_{LDR}$

$= 5 \times 10^{-5} \times 90 \times 10^3 = 4.5\,V$

b) i) NPN transistor
   ii) As it gets darker, the resistance of the LDR increases. So the voltage across the LDR increases and when it is ≥ certain value transistor (or component X) switches on and there is a current in the relay. The relay switch closes and the lamp lights.
   iii) $P = IV$

$60 = I \times 230$

$I = \dfrac{60}{230} = 0.26\,A$

Answers to Section 2

4 Heat (pages 46–47)

1 a) It requires 8360 J to change the temperature of 2.0 kg of water by 1.0 °C.
   It requires (2 × 8360) to change the temperature of (2 × 2.0) kg by 1.0 °C.
   It requires 16 720 J to change the temperature of 4.0 kg of water by 1.0 °C.
   or $E_h = cm\Delta T = 4180 \times 4 \times 1 = 16\,720\,J$

   b) It requires 16 720 J to change the temperature of 4.0 kg of water by 1.0 °C.
   It requires (5 × 16 720) J to change the temperature of 4.0 kg of water by (5 × 1) °C.
   It requires 83 600 J to change the temperature of 4.0 kg of water by 5.0 °C.
   or $E_h = cm\Delta T = 4180 \times 4 \times 5 = 83\,600\,J$

   c) It requires 83 600 J to change the temperature of 4.0 kg of water by 5.0 °C.
   It requires (2 × 83 600) J to change the temperature of (2 × 4.0) kg by 5.0 °C.
   It requires 167 200 J to change the temperature of 8.0 kg of water by 5.0 °C.
   It requires (2 × 167 200) J to change the temperature of 8.0 kg of water by (2 × 5) °C.
   It requires 334 400 J to change the temperature of 8.0 kg of water by 10 °C.
   or $E_h = cm\Delta T = 4180 \times 8 \times 10 = 334\,400\,J$

2 a) $E_h = cm\Delta T = 4180 \times 0.5 \times (58 - 18) = 83\,600\,J$
   Note: c for water from *Data Sheet*
   b) $E_h = cm\Delta T = 480 \times 0.95 \times (198 - 18) = 82\,080\,J$
   Note: c for iron from *Data Sheet*

3 Specific heat capacity is the energy required to increase the temperature of 1 kg of the material by 1 °C.
   Or 4180 J of energy is required to increase the temperature of 1 kg of water by 1 °C.

4 a) $E_h = cm\Delta T = 4180 \times 1.2 \times (100 - 20) = 401\,280\,J$
       Note: c for water from *Data Sheet*

   b) $P = \dfrac{E}{t} = \dfrac{401\,280}{180} = 2229\,W = 2.2\,kW$

5 a) $P = IV = 4 \times 12 = 48\,W$

       $P = \dfrac{E}{t}$

       $48 = \dfrac{E}{5 \times 60}$ Note: 5 minutes = $(5 \times 60)\,s$

       $E = 48 \times 300 = 14\,400\,J$

   b) Assuming no energy is transferred to the surroundings then:
       energy absorbed by copper = energy supplied = 14 400 J
       $E_h = cm\Delta T$
       $14\,400 = 386 \times 1 \times \Delta T$ Note: c for copper from *Data Sheet*

       $\Delta T = \dfrac{14\,400}{386} = 37\,°C$

   c) Some of the heat supplied by the heater will be transferred to the surroundings. The copper block will absorb less of the energy supplied and so the rise in temperature will be smaller.

6 a) $E = P \times t = 2000 \times 40 = 80\,000\,J$

   b) Assuming no energy is transferred to the surroundings then:
       energy absorbed by water = energy supplied = 80 000 J
       $E_h = cm\Delta T$
       $80\,000 = 4180 \times 0.4 \times \Delta T$ Note: c for water from *Data Sheet*

       $\Delta T = \dfrac{80\,000}{4180 \times 0.4} = 47.8\,°C$
       final temperature = initial temperature + ΔT
       $= 47.8 + 18 = 65.8\,°C$

   c) Some of the energy supplied by the heater will be transferred to the surroundings. The water will absorb less of the energy supplied by the heater and so the rise in temperature will be smaller.

7 $E_h = ml = 0.8 \times 3.34 \times 10^5 = 2.67 \times 10^5\,J$
   Note: l_{fusion} for water from *Data Sheet*

8 $E_h = ml = 0.2 \times 22.6 \times 10^5 = 4.52 \times 10^5\,J$
   Note: $l_{vaporisation}$ for water from *Data Sheet*

9 Specific latent heat of fusion is the energy required to change 1 kg from solid to liquid without change in temperature.

10 $P = \dfrac{E}{t}$

    $2000 = \dfrac{E}{80}$

    $E = 2000 \times 80 = 1.6 \times 10^5\,J$

Note: $l_{vaporisation}$ for water from *Data Sheet*
Assuming no energy is transferred to the surroundings then:
$E_h = ml$
$1.6 \times 10^5 = m \times 22.6 \times 10^5$
$m = \dfrac{1.6 \times 10^5}{2.26 \times 10^6} = 0.071\,kg$

11 energy supplied = $P \times t = 50 \times (6 \times 60) = 18\,000\,J$
    energy absorbed = $ml = (0.285 - 0.278) \times l = 0.007\,l$
    Assuming no energy is transferred to the surroundings then:
    $0.007l = 18\,000$
    $l = \dfrac{18\,000}{0.007} = 2.57 \times 10^6\,J\,kg^{-1}$

12 Let $x°C$ be the final temperature of the mixture where $20°C < x < 80°C$
    energy lost by 'hot' water = $cm\Delta T_{hot}$
    $= 4180 \times 0.1 \times (80 - x)$
    $= 418(80 - x)$
    energy gained by 'cold' water = $cm\Delta T_{cold}$
    $= 4180 \times 0.2 \times (x - 20)$
    $= 836(x - 20)$
    Assuming no energy is transferred to the surroundings then:
    energy gained by 'cold' water = energy lost by 'hot' water
    $836(x - 20) = 418(80 - x)$
    $836x - 16\,720 = 33\,440 - 418x$
    $836x + 418x = 33\,440 + 16\,720$
    $1254x = 50\,160$
    $x = \dfrac{50\,160}{1254} = 40\,°C$

13
    energy released by steam = energy released as steam changes state + energy released as condensed steam (water) changes temperature
    energy released by steam = $ml + cm\Delta T$
    energy released by steam =
    $(35 \times 10^{-3} \times 22.6 \times 10^5) + (4180 \times 35 \times 10^{-3} \times (100 - 40))$
    energy released by steam = $79\,100 + 8778$
    $= 87\,878$
    energy released by steam = $8.79 \times 10^4\,J$

5 Gas laws and the kinetic model (page 58)

1 a) i) Least pressure means largest area, 1.5 m × 0.30 m.
       ii) Greatest pressure means smallest area, 0.20 m × 0.30 m.

b) Least $p = \dfrac{F}{A} = \dfrac{100}{1.5 \times 0.3} = 222.2 = 222\,\text{Pa}$

Greatest $p = \dfrac{F}{A} = \dfrac{100}{0.2 \times 0.3} = 1666.6 = 1667\,\text{Pa}$

2    $p = \dfrac{F}{A}$

$1 \times 10^5 = \dfrac{F}{3 \times 5}$

$F = 1 \times 10^5 \times 15 = 1.5 \times 10^6\,\text{N}$

3 a) Pressure is the force exerted on 1 square metre.

  b) $p = \dfrac{F}{A}$

$250 = \dfrac{50}{A}$

$A = \dfrac{50}{250} = 0.2\,\text{m}^2$

4 a) $TK = 273 + t\,^\circ C$
    i) $273\,\text{K}$
    ii) $300\,\text{K}$
    iii) $373\,\text{K}$
    iv) $400\,\text{K}$
    v) $100\,\text{K}$
  b) $TK = 273 + t\,^\circ C$
  Therefore $t\,^\circ C = TK - 273$
    i) $-273\,^\circ C$
    ii) $-73\,^\circ C$
    iii) $20\,^\circ C$
    iv) $37\,^\circ C$
    v) $80\,^\circ C$

5     $p_1 V_1 = p_2 V_2$

$1 \times 10^5 \times 0.2 = p_2 \times 0.05$

$p_2 = \dfrac{1 \times 10^5 \times 0.2}{0.05}$

$= 400\,000 = 4.0 \times 10^5\,\text{Pa}$

6 a) Note: volume of a cylinder $\propto$ length, since area of cylinder is constant

$p_1 V_1 = p_2 V_2$

$1 \times 400 = 8 \times V_2$

$V_2 = \dfrac{1 \times 400}{8} = 50\,\text{mm}$

Distance pushed in $= 400 - 50 = 350\,\text{mm}$

  b) Temperature of gas remained constant.

7     $\dfrac{V_1}{T_1} = \dfrac{V_2}{T_2}$

$\dfrac{1600}{17 + 273} = \dfrac{2000}{T_2}$ Note: temperatures must be in Kelvin

$T_2 = \dfrac{2000 \times 290}{1600} = 363\,\text{K}$

8     $\dfrac{V_1}{T_1} = \dfrac{V_2}{T_2}$

$\dfrac{5}{20 + 273} = \dfrac{V_2}{-18 + 273}$ Note: temperatures must be in Kelvin

$V_2 = \dfrac{5 \times 255}{293} = 4.35\,\text{litres}$

9 Note: $75\,\text{kPa} = 75 \times 10^3\,\text{Pa} = 75\,000\,\text{Pa}$

$\dfrac{p_1}{T_1} = \dfrac{p_2}{T_2}$

$\dfrac{75 \times 10^3}{27 + 273} = \dfrac{p_2}{37 + 273}$ Note: temperatures must be in Kelvin

$p_2 = \dfrac{75 \times 10^3 \times 310}{300} = 77\,500\,\text{Pa} = 77.5\,\text{kPa}$

10     $\dfrac{p_1}{T_1} = \dfrac{p_2}{T_2}$

$\dfrac{2.5 \times 10^5}{T_1} = \dfrac{3.0 \times 10^5}{27 + 273}$ Note: temperatures must be in Kelvin

$T_1 = \dfrac{2.5 \times 10^5 \times 300}{3.0 \times 10^5}$

$= 250\,\text{K} \,(= 250 - 273 = -23\,^\circ C)$

11     $\dfrac{p_1 V_1}{T_1} = \dfrac{p_2 V_2}{T_2}$

$\dfrac{1.0 \times 10^5 \times 100}{300} = \dfrac{p_2 \times 200}{400}$ Note: temperatures are in Kelvin

$33\,333 = p_2 \times 0.5$

$p_2 = \dfrac{33\,333}{0.5} = 6.7 \times 10^4\,\text{Pa}$

12 a) As the temperature of the gas falls, the average kinetic energy of the particles decreases, so the particles have less speed. Particles hit the walls of the container with a smaller force and less often (or fewer times per second). This means there is less particle bombardment on the walls, so the pressure exerted by the gas is less.

  b) Particles stop moving.

Exam practice for Chapters 1–5 (pages 59-61)

1 a) i) $R_T = R_1 + R_2 = 36 + 18 = 54\,\Omega$

$V_{supply} = I_{circuit} R_T$

$9 = I_{circuit} \times 54$

$I_{circuit} = \dfrac{9}{54} = 0.167\,\text{A}$

    ii) $V_{18\Omega} = (IR)_{18\Omega} = I_{circuit} R_{18\Omega} = 0.167 \times 18 = 3\,\text{V}$

b) i) $\dfrac{1}{R_{XY}} = \dfrac{1}{R_1} + \dfrac{1}{R_2} + \dfrac{1}{R_3} = \dfrac{1}{100} + \dfrac{1}{75} + \dfrac{1}{100}$

$\dfrac{1}{R_{XY}} = 0.010 + 0.013 + 0.010 = 0.033$

$R_{XY} = \dfrac{1}{0.033} = 30\,\Omega$

Note: for parallel resistors, answer must be smaller than smallest R

ii) $P = \dfrac{V^2}{R_{XY}} = \dfrac{230^2}{30} = 1763\,W = 1.76 \times 10^3\,W$

$= 1.76\,kW$

2 a) To ensure correct voltage is applied to each lamp

b) $P = IV$

$3 = I \times 12$

$I = \dfrac{3}{12} = 0.25\,A$

c) Since lamps and resistor are connected in series, then:

$V_{supply} = (19 \times V_{lamp}) + V_{resistor}$

$230 = (19 \times 12) + V_{resistor}$

$230 = 228 + V_{resistor}$

$V_{resistor} = 2.0\,V$

But

voltage across resistor $=$ current in resistor $\times$ resistance of resistor

$V_{resistor} = (IR)_{resistor}$

$2 = 0.25 \times R$

$R = \dfrac{2}{0.25} = 8.0\,\Omega$

d) $Q = It = 0.25 \times 60$ Note: 1 minute = 60 s

$Q = 15\,C$

e) voltage across lamp $=$ current in lamp $\times$ resistance of lamp

$V_{lamp} = (IR)_{lamp}$

$12 = 0.25 \times R_{lamp}$

$R_{lamp} = \dfrac{12}{0.25} = 48\,\Omega$

3 a) $P = \dfrac{E}{t}$

$2116 = \dfrac{E}{180}$

$E = 2116 \times 180 = 3.81 \times 10^5\,J$

b) $P = \dfrac{V^2}{R}$

$2116 = \dfrac{230^2}{R}$

$R = \dfrac{52\,900}{2116} = 25\,\Omega$

4 a) LED

b) To prevent damage to the LED from too high a current in the LED (or too high a voltage across the LED)

c) Since components are connected in series then:

$V_{supply} = V_{LED} + V_{resistor}$

$10 = 1.8 + V_{resistor}$

$V_{resistor} = 8.2\,V$

But: voltage across resistor $=$ current in resistor $\times$ resistance of resistor

$V_{resistor} = (IR)_{resistor}$

Note: current is the same in a series circuit

$8.2 = 11 \times 10^{-3} \times R$

$R = \dfrac{8.2}{11 \times 10^{-3}} = 745\,\Omega$

5 a) Circuit X

b) i) $R_T = R_1 + R_2 = 100 + 100 = 200\,\Omega$

ii) $\dfrac{1}{R_T} = \dfrac{1}{R_1} + \dfrac{1}{R_2}$

$= \dfrac{1}{100} + \dfrac{1}{100}$

$= 0.010 + 0.010 = 0.020$

$\dfrac{1}{R_T} = 0.020$

$R_T = \dfrac{1}{0.020} = 50\,\Omega$

c) i) $V = IR$

$230 = I \times 200$

$I = \dfrac{230}{200} = 1.15\,A$

ii) $V = IR$

$230 = I \times 50$

$I = \dfrac{230}{50} = 4.6\,A$

d) Circuit Y – it has the higher power rating since lower resistance means higher current for the same voltage (or by calculation).

6 a) n-channel enhancement MOSFET

b) As the temperature of the thermistor increases, the resistance of the thermistor decreases and so the voltage across the thermistor decreases. The voltage across the variable resistor must therefore increase and when it is equal to or greater than a certain value the MOSFET switches on and the LED lights.

c) Change the positions of the thermistor and the variable resistor.

7 a) Y = MOSFET transistor; Z = relay

b) Resistance of LDR increases as light level falls.

c) As the light level falls, resistance of LDR increases, so voltage across LDR increases and when it ≥ 2.6 V, component Y (or MOSFET transistor) switches on. The resulting current in the relay coil closes the switch lighting the lamp.

d) To adjust the light level at which the lamp lights.

8 Note: c_{water}, c_{ice} and l_{fusion} for water from *Data Sheet*
Water at 20 °C to 0 °C:
$E_h = cm\Delta T = 4180 \times 0.15 \times 20 = 12\,540\,J$
Water at 0 °C changes to ice at 0 °C:
$E_h = ml = 0.15 \times 3.34 \times 10^5 = 50\,100\,J$
Ice at 0 °C to −19 °C:
$E_h = cm\Delta T = 2100 \times 0.15 \times 19 = 5985\,J$
Total energy removed = 12 540 + 50 100 + 5985
$$= 68\,625 = 68.6\,kJ$$

9 a) $E_h = cm\Delta T = 4180 \times 1.2 \times 12.5 = 62\,700\,J$
Note: c_{water} from *Data Sheet*

b) i) $P = \dfrac{E}{t} = \dfrac{62\,700}{23 \times 60} = 45.4\,W$
Note: 23 minutes = (23 × 60) s
ii) Assuming that all the energy supplied by the heater is absorbed by the water

c) $P = IV$
$45.4 = I \times 12$
$I = \dfrac{45.4}{12} = 3.8\,A$

10 a) 80 °C

b) Note: ΔT of (80 − 20) °C occurs between 0 s and 330 s
Assuming that all the energy supplied by the heater is absorbed by the sample then:
$$E_h = cm\Delta T$$
$$P \times t = cm\Delta T$$
$$(100 \times 330) = c \times 0.4 \times (80 - 20)$$
$$c = \frac{33\,000}{24} = 1375\,J\,kg^{-1}\,°C^{-1}$$

c) Note: change of state between 330 s and 830 s
Assuming that all the energy supplied by the heater is absorbed by the sample changing state then:
$$E_h = ml$$
$$P \times t = ml$$
$$100 \times (830 - 330) = 0.4 \times l$$
$$l = \frac{50\,000}{0.4} = 125\,000\,J\,kg^{-1}$$

11 a) $p_1 V_1 = p_2 V_2$
$p_1 \times 7 = 100 \times 10^3 \times (3.5 + 7)$ Note: $V_2 = V_X + V_Y$
$p_1 = \dfrac{100 \times 10^3 \times 10.5}{7} = 150\,000\,Pa = 150\,kPa$

b) The temperature of the air did not change.

12 $\dfrac{V_1}{T_1} = \dfrac{V_2}{T_2}$

$\dfrac{500}{20 + 273} = \dfrac{V_2}{5 + 273}$ Note: temperatures must be in Kelvin

$\dfrac{500}{293} = \dfrac{V_2}{278}$

$V_2 = \dfrac{500 \times 278}{293} = 474\,m^3$

13 $\dfrac{p_1}{T_1} = \dfrac{p_2}{T_2}$

$\dfrac{1.5 \times 10^5}{10 + 273} = \dfrac{2.2 \times 10^5}{T_2}$ Note: temperatures must be in Kelvin

$\dfrac{1.5 \times 10^5}{283} = \dfrac{2.2 \times 10^5}{T_2}$

$T_2 = \dfrac{2.2 \times 10^5 \times 283}{1.5 \times 10^5} = 415\,K\,(= 142\,°C)$

14 $\dfrac{V_1}{T_1} = \dfrac{V_2}{T_2}$

$\dfrac{0.0025}{-70 + 273} = \dfrac{V_2}{27 + 273}$ Note: temperatures must be in Kelvin

$\dfrac{0.0025}{203} = \dfrac{V_2}{300}$

$V_2 = \dfrac{0.0025 \times 300}{203} = 0.0037\,m^3$

Answers to Section 3

6 Waves and wave phenomena (pages 75-77)

1 a) Waveform X

b) Waveform X since gamma is a member of the electromagnetic spectrum, all of which are transverse waves.

2 a) $f = \dfrac{1}{T} = \dfrac{1}{0.2} = 5.0\,\text{Hz}$

b) $v = f\lambda = 5 \times 3 = 15\,\text{m s}^{-1}$

c) $T = \dfrac{1}{f} = \dfrac{1}{10} = 0.1\,\text{s}$

d) $v = f\lambda$

$25 = 10 \times \lambda$

$\lambda = \dfrac{25}{10} = 2.5\,\text{m}$

e) $f = \dfrac{1}{T} = \dfrac{1}{0.125} = 8.0\,\text{Hz}$

f) $v = f\lambda = 8 \times 4 = 32\,\text{m s}^{-1}$

g) $f = \dfrac{1}{T} = \dfrac{1}{2.5} = 0.4\,\text{Hz}$

h) Note: $800\,\text{mm s}^{-1} = 800 \times 10^{-3}\,\text{m s}^{-1}$

$v = f\lambda$

$800 \times 10^{-3} = 0.4 \times \lambda$

$\lambda = \dfrac{800 \times 10^{-3}}{0.4} = 2.0\,\text{m}\ (= 2000\,\text{mm})$

3 a) $2A = 0.5$

$A = \dfrac{0.5}{2} = 0.25\,\text{m}$

b) Note: 4 waves between X and Y

$\text{frequency} = \dfrac{\text{number of waves}}{\text{time taken}} = \dfrac{4}{0.2} = 20\,\text{Hz}$

c) $4\lambda = 2$

$\lambda = \dfrac{2}{4} = 0.5\,\text{m}$

d) $v = f\lambda = 20 \times 0.5 = 10\,\text{ms}^{-1}$ or $v = \dfrac{d}{t} = \dfrac{2}{0.2} = 10\,\text{ms}^{-1}$

4 a) $v = \dfrac{d}{t} = \dfrac{30}{20} = 1.5\,\text{m s}^{-1}$

b) $v = f\lambda$

$1.5 = f \times 3$

$f = \dfrac{1.5}{3} = 0.5\,\text{Hz}$

c) The speed of the waves is unchanged as the medium is unchanged. When frequency is doubled then wavelength is halved (to keep speed the same).

5 a) Refraction is the change in speed of light as it passes from one medium to another.

b) and **c)**

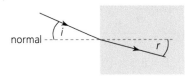

6 A = slows down; B = bends towards; C = refraction; D = speeds up; E = bends away

7 a) Angle of incidence = 30°

b) Light slows down in the glass and the refracted ray bends towards the normal.

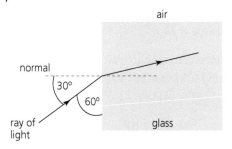

8 a) and **b)** Light speeds up in going from glass into water and the ray bends away from the normal.

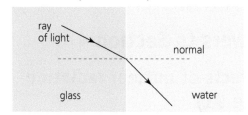

9 a) Slows down (as ray of light bends towards the normal)

b) i) 50°

ii) 30°

10

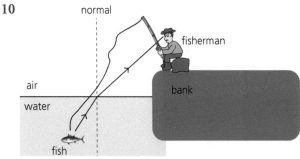

11 Radio waves with the longer wavelength (lower frequency) will be diffracted round the hills the most. The man should tune in to the LW waveband.

12 a) i) P = infrared, Q = X-rays

ii) Photographic film

b) They are transverse waves.

They travel at a speed of $3 \times 10^8\,\text{m s}^{-1}$ in air (and in a vacuum).

They all transfer energy.

13 a) Note: electromagnetic waves travel at $3 \times 10^8\,\text{m s}^{-1}$

$v = f\lambda$

$3 \times 10^8 = f \times 4.80 \times 10^{-7}$

$f = \dfrac{3 \times 10^8}{4.80 \times 10^{-7}} = 6.25 \times 10^{14}\,\text{Hz}$

b) The speed of blue light is the same as that of red light ($3 \times 10^8 \, \text{m s}^{-1}$). Since red light has a longer wavelength than blue light, the frequency of red light must be smaller than that of blue light (v same $= f\!\downarrow \lambda\!\uparrow$).

14 Note: electromagnetic waves travel at $3 \times 10^8 \, \text{m s}^{-1}$

Note: $384\,400 \, \text{km} = 384\,400 \times 10^3 \, \text{m} = 384\,400\,000 \, \text{m}$

$$v = \frac{d}{t}$$

$$3 \times 10^8 = \frac{384\,400 \times 10^3}{t}$$

$$t = \frac{384\,400 \times 10^3}{3 \times 10^8}$$

$$t = 1.28 \, \text{s}$$

Answers to Section 4

7 Effects of nuclear radiation (page 91)

1

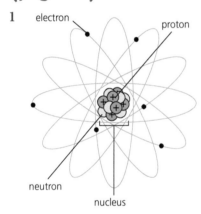

or

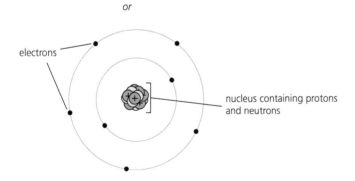

Protons are positively charged, neutrons are neutral and electrons are negatively charged.

2 a) An alpha particle is a helium nucleus and consists of two protons and two neutrons. It is positively charged.

b) A beta particle is a high-speed electron. It is negatively charged.

3 a) Ionisation is the breaking up of a neutral atom into positive and negative pieces.

b) Alpha radiation

c) Gamma radiation

4 a) To check on the amount and type of radiation the wearer has been exposed to.

b) The open window and 0.1 mm plastic will be affected. (β particles are absorbed by a few mm of aluminium. The 3 mm aluminium may or may not be affected – it depends on the strength of beta radiation. The 1 mm lead will be unaffected.)

5 A gamma source, because it will be able to penetrate through the body and so be detected outside the patient's body. Alpha and beta radiation will be absorbed by the body.

6 Place each source, in turn, close to the GM tube. Place thin paper between the source and the GM tube. For the gamma and beta sources, the count rate will be unaffected by the paper *or* for the alpha source, the count rate will fall almost to zero. Place the remaining two sources, in turn, close to the GM tube. Place 5 mm thick aluminium between the source and the tube. For the gamma source, the count rate will be unaffected by the aluminium *or* for the beta source, the count rate will fall almost to zero.

7 The absorbed dose, the type of radiation absorbed and the tissue or the body organs that are exposed.

8 Note: $15 \, \mu\text{J} = 15 \times 10^{-6} \, \text{J} = 0.000\,015 \, \text{J}$

$$D = \frac{E}{m} = \frac{15 \times 10^{-6}}{0.25} = 60 \times 10^{-6} \, \text{Gy}$$

$$D = 6.0 \times 10^{-5} \, \text{Gy}$$

9 a) Absorbed dose is the energy absorbed by 1 kg of the absorbing material.

b) Note: $8.5 \, \text{mGy} = 8.5 \times 10^{-3} \, \text{Gy} = 0.0085 \, \text{Gy}$
and $20 \, \text{g} = 20 \times 10^{-3} \, \text{kg} = 0.020 \, \text{kg}$

$$D = \frac{E}{m}$$

$$8.5 \times 10^{-3} = \frac{E}{20 \times 10^{-3}}$$

$$E = 8.5 \times 10^{-3} \times 20 \times 10^{-3} = 1.7 \times 10^{-4} \, \text{J}$$

c) $H = D \times w_R$ Note: w_R from *Data Sheet*
$H = 8.5 \times 10^{-3} \times 20 = 0.17 \, \text{Sv}$

10 a) The radiation weighting factor takes account of the damage done by different kinds of radiation.

b) Radiation weighting factor for fast neutrons = 10
Note: w_R from *Data Sheet*

11 Note: $125\,g = 125 \times 10^{-3}\,kg = 0.125\,kg$
$4.5\,mJ = 4.5 \times 10^{-3}\,J = 0.0045\,J$
$$D = \frac{E}{m} = \frac{4.5 \times 10^{-3}}{0.125 \times 10^{-3}} = 0.036\,Gy$$
$H = D \times w_R = 0.036 \times 1$ Note: w_R from *Data Sheet*
$H = 0.036\,Sv$

12 a) Note: w_R values from *Data Sheet*
Note: $75\,\mu Gy = 75 \times 10^{-6}\,Gy$
and $15\,mGy = 1.5 \times 10^{-3}\,Gy$
$$H_\alpha = D \times w_R = 75 \times 10^{-6} \times 20 = 1.5 \times 10^{-3}\,Sv$$
$$H_{slow} = D \times w_R = 15 \times 10^{-3} \times 3 = 45 \times 10^{-3}\,Sv$$
$$\begin{aligned} H_{total} &= H_{slow} + H_{fast} \\ &= 1.5 \times 10^{-3} + 45 \times 10^{-3} \\ &= 0.047\,Sv \end{aligned}$$

b) Any two from:
Shield tissue from radiation, e.g. place a piece of paper between source and the tissue.
Move the tissue further away from the source of radiation.
Limit the time of exposure to the radiation.

13 Note: $200\,\mu Sv = 200 \times 10^{-6}\,Sv = 0.000\,200\,Sv$
$H = D \times w_R$ Note: w_R from *Data Sheet*
$200 \times 10^{-6} = D \times 20$
$$D = \frac{200 \times 10^{-6}}{20} = 10 \times 10^{-6}\,Gy\ (= 1 \times 10^{-5}\,Gy)$$

14 a) Note: $1.17\,mSv = 1.17 \times 10^{-3}\,Sv = 0.001\,17\,Sv$
and $117\,\mu Gy = 117 \times 10^{-6}\,Gy = 0.000\,117\,Gy$
$$H = D \times w_R$$
$$1.17 \times 10^{-3} = 117 \times 10^{-6} \times w_R$$
$$w_R = \frac{1.17 \times 10^{-3}}{117 \times 10^{-6}} = 10$$

b) From *Data Sheet*, an w_R of 10 is fast neutrons.

15 Note: $210\,\mu Sv\,h^{-1} = 210 \times 10^{-6}\,Sv\,h^{-1} = 0.000\,210\,Sv\,h^{-1}$
$$\dot{H} = \frac{H}{t}$$
$$210 \times 10^{-6} = \frac{H}{8}$$
$$H = 210 \times 10^{-6} \times 8 = 1.68 \times 10^{-3}\,Sv$$

8 Using the nucleus (page 99)

1 a) 3.5 MBq means that 3.5 million radioactive nuclei decay (disintegrate) every second.

b) Any two from:

Always use a lifting tool or wear special gloves when moving a source; arrange the source so that the radiation window points away from the body; never point the source at your eyes; wash your hands after using radioactive sources; minimise time working with source; shield yourself from source; work as far away as possible from source.

2 a) It takes 5.3 years for half of the radioactive nuclei present to disintegrate *or* it takes 5.3 years for the activity of the radioactive source to halve.

b)

Number of half-lives	Activity	Time (years)
0	1	0
1	$\frac{1}{2}$	5.3
2	$\frac{1}{4}$	10.6
3	$\frac{1}{8}$	15.9
4	$\frac{1}{16}$	21.2

4 half-lives $= 4 \times 5.3 = 21.2$ years

3

Number of half-lives	Activity (kBq)	Time (hours)
0	1600	0
1	800	6
2	400	12
3	200	18

Activity after 18 hours is 200 kBq.

4

Number of half-lives	Activity (µBq)	Time (years)
0	2400	0
1	1200	5700
2	600	11 400
3	300	17 100
4	150	22 800

4 half-lives $= 4 \times 5700 = 22\,800$ years

5 Corrected count rate after 1 half-life is 800 kBq. From graph, 800 kBq gives a time of 5 hours.

6 Nuclear fusion is the joining together of two light nuclei to form a heavier nucleus.

Exam practice for Chapters 6–8 (pages 100-102)

1 a) i) sound
   ii) energy
   b) i) $v = \dfrac{d}{t} = \dfrac{6}{12} = 0.5\,\text{m s}^{-1}$
   ii) frequency $= \dfrac{N}{t} = \dfrac{20}{60} = 0.33\,\text{Hz}$
   $v = f\lambda$
   $0.5 = 0.33 \times \lambda$
   $\lambda = \dfrac{0.5}{0.33} = 1.5\,\text{m}$

2 a) i) P = microwaves, Q = ultraviolet
   ii) Photographic film
   b) Since speed of these waves is constant at $3 \times 10^8\,\text{m s}^{-1}$, then longest wavelength will have the lowest frequency.
   $v = f\lambda$
   $3 \times 10^8 = 1.8 \times 10^{10} \times \lambda$
   $\lambda = \dfrac{3 \times 10^8}{1.8 \times 10^{10}} = 0.017\,\text{m}$

3 a) i) 40°
   ii) 25°
   b) i) Note: electromagnetic waves travel at $3 \times 10^8\,\text{m s}^{-1}$
   $v = f\lambda$
   $3 \times 10^8 = 4.8 \times 10^{14} \times \lambda$
   $\lambda = \dfrac{3 \times 10^8}{4.8 \times 10^{14}} = 6.3 \times 10^{-7}\,\text{m}$
   ii) $4.8 \times 10^{14}\,\text{Hz}$ since frequency of a wave is determined by the source.

4 a) Diffraction is the bending of waves at a gap or obstacle.
   b)

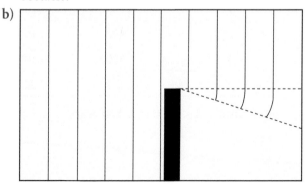

c) The speed of the waves in the tank remains the same so increasing the frequency will decrease the wavelength of the waves. Smaller wavelength will decrease the amount of diffraction.

5 a) Alpha particles would not pass through the cardboard and so would not be detected by the GM tube.
   Gamma rays are too penetrating and any minor change in thickness of the cardboard would not make any difference to the reading on the counter.
   b) The reading would increase.
   c) Any two from:
   Always use a lifting tool or wear special gloves when moving a source; arrange the source so that the radiation window points away from the body; never point the source at your eyes; wash your hands after using radioactive sources; minimise time working with the source; shield yourself from the source; work as far away as possible from the source.

6 a) Ionisation is the breaking up of a neutral atom into positive and negative pieces.
   b) Alpha source, since the alpha source would produce the most ionisation of the three sources and would be more easily absorbed by the smoke particles.

7 a) Source X
   b) The radiation from source Y would be absorbed by the patient's body.
   Source Z has too long a half-life so the level of radiation would be too high for a long time. This would mean unnecessary damage would be done to the patient.
   c) Alpha radiation would be absorbed by the patient's body and so would not be detected outside the body, and alpha radiation produces a lot of ionisation, which damages body tissue.

8 a) Gamma radiation is highly penetrating and is able to pass out of the body, and so can be detected.
   b) i) The half-life of a radioactive source is the time taken for half of the radioactive nuclei present to disintegrate.
   or
   It is the time taken for the activity of the radioactive source to halve.

ii)

Number of half-lives	Activity (MBq)	Time (hours)
0	640	0
1	320	6
2	160	12
3	80	18
4	40	24

Activity after 24 hours is 40 MBq.

c) Any two from: energy of the radiation, type of radiation, type of living material.

9 a) Alpha particles are helium nuclei, which consist of two protons and two neutrons and are positively charged.

b) 20 counts per minute

c) $\dfrac{\text{total count}}{\text{rate}} = \dfrac{\text{source count}}{\text{rate}} + \dfrac{\text{background count}}{\text{rate}}$

When total count rate = 180 cpm, source count rate = 160 cpm.

For half-life, source count rate is required to fall by half, i.e. new source count rate = 80 cpm. This gives a total count rate of 80 + 20 = 100 cpm. From the graph, a total count rate of 100 cpm gives a time of 1 minute.

Half-life = 1 minute

10 a) Fission is the splitting up of a heavy nucleus into two lighter nuclei with the emission of two or three neutrons – in this case a uranium nucleus splits into two pieces and two or three neutrons are released.

b) These neutrons are used to produce further fissions and keep the reaction going.

c) $H_{\text{total}} = H_{\text{fast}} + H_{\text{slow}} + H_{\gamma}$
$H_{\text{total}} = (D \times w_R)_{\text{fast}} + (D \times w_R)_{\text{slow}} + (D \times w_R)_{\gamma}$
$H_{\text{total}} = (3 \times 10^{-6} \times 10) + (2 \times 10^{-6} \times 3) +$
$\qquad\qquad (1.5 \times 10^{-6} \times 1)$
$H_{\text{total}} = (30 \times 10^{-6}) + (6 \times 10^{-6}) + 1.5 \times 10^{-6}$
$H_{\text{total}} = 37.5 \times 10^{-6}\,\text{Sv}\ (= 37.5\,\mu\text{Sv})$

Answers to Section 5

9 Kinematics (page 115)

1 average speed $= \dfrac{d}{t} = \dfrac{4.2}{3.5} = 1.2\,\text{m s}^{-1}$

2 average speed $= \dfrac{d}{t}$

$60 = \dfrac{d}{0.5}$ Note: 30 minutes = 0.5 hours

$d = 60 \times 0.5 = 30\,\text{km}$

3 average speed $= \dfrac{d}{t}$

$6.0 = \dfrac{400}{t}$

$t = \dfrac{400}{6.0} = 66.7 = 67\,\text{s}$

4 Velocity is the distance travelled in a certain direction in 1 second.

5 a) distance travelled = 300 + 400 = 700 m

b) Using a vector diagram:

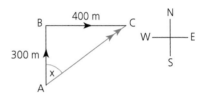

Resultant displacement = AC
$AC^2 = AB^2 + BC^2 = 300^2 + 400^2 = 250\,000$
$AC = 500\,\text{m}$
$\tan x = \dfrac{BC}{AB} = \dfrac{400}{300} = 1.33$
$x = 53°$
displacement = 500 m at 53 E of N
If using a scale vector diagram, then
500 ± 10 m at 53 ± 2° E of N

c) Note: 7 minutes = (7 × 60) s
average velocity $= \dfrac{\text{displacement}}{\text{time taken}}$
$= \dfrac{500\,\text{m}}{(7 \times 60)\,\text{s}}$ at 53° E of N
average velocity = 1.19 m s⁻¹ at 53° E of N

6 Using a vector diagram:

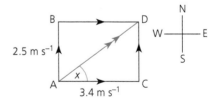

resultant velocity = AD
$AD^2 = AC^2 + CD^2 = 3.4^2 + 2.5^2 = 11.56 + 6.25 = 17.81$
$AD = \sqrt{17.81} = 4.2\,\text{m s}^{-1}$
$\tan x = \dfrac{CD}{AC} = \dfrac{2.5}{3.4} = 0.735$
$x = 36.3°$
resultant velocity of boat = 4.2 m s⁻¹ at 36° N of E
If using a scale vector diagram, then 4.2 ± 0.2 m s⁻¹ at 36 ± 2° N of E

7 $5.0\,\mathrm{m\,s^{-2}}$ means that the velocity of the object increases by $5.0\,\mathrm{m\,s^{-1}}$ per second.

8 a) $a = \dfrac{v - u}{t} = \dfrac{4 - 0}{8} = \dfrac{4}{8} = 0.5\,\mathrm{m\,s^{-2}}$

   b) $a = \dfrac{v - u}{t}$

   $0.5 = \dfrac{6 - 0}{t}$

   $0.5 = \dfrac{6}{t}$

   $t = \dfrac{6}{0.5} = 12\,\mathrm{s}$

9 $a = \dfrac{v - u}{t}$

   $0.2 = \dfrac{1.5 - u}{5}$

   $0.2 \times 5 = 1.5 - u$

   $1 = 1.5 - u$

   $u = 1.5 - 1 = 0.5\,\mathrm{m\,s^{-1}}$

10 $a = \dfrac{v - u}{t}$

   $-1.5 = \dfrac{v - 30}{18}$

   $-1.5 \times 18 = v - 30$

   $-27 = v - 30$

   $v = 30 - 27 = 3\,\mathrm{m\,s^{-1}}$

11 $a = \dfrac{v - u}{t}$

   $5 = \dfrac{12\,000 - 0}{t}$

   $t = \dfrac{12\,000}{5} = 2400\,\mathrm{s}$

12 a) $a = \dfrac{v - u}{t} = \dfrac{12 - 0}{4} = \dfrac{12}{4} = 3\,\mathrm{m\,s^{-2}}$

   b) average velocity $= \dfrac{u + v}{2}$

   Note: this equation may be used because acceleration is constant.

   average velocity $= \dfrac{u + v}{2} = \dfrac{0 + 12}{2} = 6\,\mathrm{m\,s^{-1}}$

13 a) Constant negative acceleration (or constant deceleration) from $10\,\mathrm{m\,s^{-1}}$ to $8\,\mathrm{m\,s^{-1}}$ in $12\,\mathrm{s}$.

   b) Note: $u = 10\,\mathrm{m\,s^{-1}}$ and $v = 8\,\mathrm{m\,s^{-1}}$

   $a = \dfrac{v - u}{t} = \dfrac{8 - 10}{12} = \dfrac{-2}{12} = -0.17\,\mathrm{m\,s^{-2}}$

   c) average velocity $= \dfrac{u + v}{2} = \dfrac{10 + 8}{2} = 9\,\mathrm{m\,s^{-1}}$

   Note: this equation may be used because acceleration is constant.

14 a) i) Constant acceleration from $5\,\mathrm{m\,s^{-1}}$ to $20\,\mathrm{m\,s^{-1}}$ in $100\,\mathrm{s}$.

   ii) Constant velocity of $20\,\mathrm{m\,s^{-1}}$ for $200\,\mathrm{s}$.

   b) $a = \dfrac{v - u}{t} = \dfrac{20 - 5}{100} = \dfrac{15}{100} = 0.15\,\mathrm{m\,s^{-2}}$

   c) displacement = area under velocity–time graph

   displacement = area of I + area of II + area of III

   $= (100 \times 5) + \left(\dfrac{1}{2} \times 100 \times 15\right) + (200 \times 20)$

   displacement $= 500 + 750 + 4000 = 5250\,\mathrm{m}$

   d) average velocity $= \dfrac{\text{displacement}}{\text{time}} = \dfrac{5250}{300} = 17.5\,\mathrm{m\,s^{-1}}$

10 Forces and their effects (pages 131-132)

1 The resultant force, represented by the line AD, can be found using either a vector diagram or a scale vector diagram.

Using a vector diagram:

$AD^2 = AC^2 + CD^2 = 500^2 + 200^2$

$AD^2 = 250\,000 + 40\,000 = 290\,000$

$AD = \sqrt{290\,000} = 538.5\,\mathrm{N}$

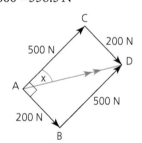

$\tan x = \dfrac{CD}{AC} = \dfrac{200}{500} = 0.4$

$x = 21.8°$

resultant force = $539\,\mathrm{N}$ at $22°$ from the $500\,\mathrm{N}$ force

Using a scale vector diagram:

Scale: $1\,\mathrm{cm} \equiv 100\,\mathrm{N}$

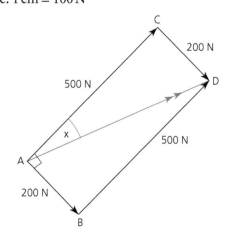

AD = 5.3 cm

1 cm ≡ 100 N, so

AD = 5.3 cm × 100 = 530 N

$x = 22°$ (using a protractor)

resultant force = 530 N at 22° from 500 N force

2 a)

direction of travel

resistive force [ ] engine force

b) i) The forces are balanced *or* the engine force is equal in size but acts in the opposite direction from the resistive force.

ii) The engine force is larger than the resistive force.

3 a) 170 kg (the number and type of particles making up the probe have not changed, so mass remains the same)

b) i) $W_{Earth} = mg = 170 × 9.8 = 1666$ N

Note: g for Earth from *Data Sheet*

ii) $W_{Mars} = mg = 170 × 3.7 = 629$ N

Note: g for Mars from *Data Sheet*

4 a) A = air resistance, B = lift force (from wings), C = engine force, D = weight

b) i) A and C are equal in size but opposite in direction

ii) B and D are equal in size but opposite in direction.

5 Gravitational field strength is the weight of a 1 kg mass.

6 a) $W = mg = 1.5 × 9.8 = 14.7$ N

Note: g for Earth from *Data Sheet*

b) Since lampshade is hanging at rest, the forces must be balanced

weight (downwards) = tension (upwards)

tension = 14.7 N upwards

7 $F_{un} = ma = 5000 × 0.12 = 600$ N

8 Note: 400 g = 400 × 10⁻³ kg = 0.400 kg

$F_{un} = ma$

$0.2 = 400 × 10^{-3} × a$

$a = \dfrac{0.2}{400 × 10^{-3}} = 0.5$ m s⁻²

9 $F_{un} = ma$

$5.4 × 10^3 = m × 2.7$ Note: 5.4 kN = 5.4 × 10³ N = 5400 N

$m = \dfrac{5.4 × 10^3}{2.7} = 2000$ kg

10 a) $F_{un} = ma = 0.3 × 4 = 1.2$ N

b)

direction of travel

resistive force [ ] applied force +

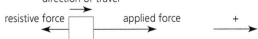

F_{un} = applied force + (−resistive force)

Note: resistive force acts in opposite direction from applied force.

1.2 = 2.0 − resistive force

resistive force = 2.0 − 1.2 = 0.8 N

11 a) $W = mg = 1.3 × 10^4 × 9.8 = 1.27 × 10^5$ N

Note: g for Earth from *Data Sheet*

b)

upward force

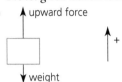

weight

c) unbalanced force = upward force + (−weight)

= $1.6 × 10^5 + (−1.27 × 10^5)$

= $3.3 × 10^4$ N

$F_{un} = ma$

$3.3 × 10^4 = 1.3 × 10^4 × a$

$a = \dfrac{3.3 × 10^4}{1.3 × 10^4} = 2.5$ m s⁻²

12 a)

direction of travel

150 N [ ] 500 N +

unbalanced force on car = 500 + (−150) = 350 N

$F_{un} = ma$

$350 = 1100 × a$

$a = \dfrac{350}{1100} = 0.32$ m s⁻²

b) Since the car is now moving at constant speed (Newton's first law) the forces acting on the car must be balanced, i.e. the resistive force has increased to 500 N and is equal in size but opposite in direction from the forward force of 500 N.

13 a) Note: 200 g = 200 × 10⁻³ kg = 0.200 kg

$W = mg = 200 × 10^{-3} × 9.8 = 1.96$ N

b) $a = \dfrac{1.96}{200 × 10^{-3}} = 9.8$ m s⁻²

14 a) horizontal speed = $\dfrac{\text{horizontal distance}}{\text{time taken}}$

Note: horizontal motion is constant speed of 18.0 m s⁻¹

$18 = \dfrac{\text{horizontal distance}}{3}$

horizontal distance = 18 × 3 = 54 m

b) $a = \dfrac{v - u}{t}$ Note: vertical motion is constant acceleration of $9.8\,\mathrm{m\,s^{-2}}$ from rest

$9.8 = \dfrac{v - 0}{3}$

$v = 9.8 \times 3 = 29.4\,\mathrm{m\,s^{-1}}$

c) height = vertical distance travelled by ball

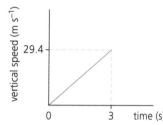

height = area under vertical speed–time graph

height = area under graph $= \dfrac{1}{2} \times 3 \times 29.4 = 44.1\,\mathrm{m}$

Alternatively:

average vertical speed $= \dfrac{u + v}{2}$

average vertical speed $= \dfrac{0 + 29.4}{2} = 14.7\,\mathrm{m\,s^{-1}}$

Note: this equation may be used because acceleration is constant.

height = vertical distance travelled

height = average vertical speed × time taken

height = 14.7×3

height = $44.1\,\mathrm{m}$

11 Newton's third law and energy (pages 143–145)

1 A = exhaust gases; B = rocket; C = third

2 Note: 16 strips of lawn each $4.0\,\mathrm{m}$ long gives

$d = (16 \times 4)\,\mathrm{m}$

$E_W = F \times d = 35 \times (16 \times 4) = 2240\,\mathrm{J}$

3 $E_W = F \times d$

$500 = 25 \times d$

$d = \dfrac{500}{25} = 20\,\mathrm{m}$

4 Note: $1.5\,\mathrm{MJ} = 1.5 \times 10^6\,\mathrm{J} = 1\,500\,000\,\mathrm{J}$

$E_W = F \times d$

$1.5 \times 10^6 = F \times 100$

$F = \dfrac{1.5 \times 10^6}{100} = 1.5 \times 10^4\,\mathrm{N}$

5 Note: $750\,\mathrm{g} = 750 \times 10^{-3}\,\mathrm{kg} = 0.750\,\mathrm{kg}$

and $350\,\mathrm{mm} = 350 \times 10^{-3}\,\mathrm{m} = 0.350\,\mathrm{m}$

change in $E_p = mgh$ Note: g from *Data Sheet*

change in $E_p = 750 \times 10^{-3} \times 9.8 \times 350 \times 10^{-3}$

change in $E_p = 2.5725\,\mathrm{J} = 2.57\,\mathrm{J}$

Note: do not put too many figures in answer

6 change in gravitational $E_p = mgh$

$2450 = 350 \times 9.8 \times h$

$h = \dfrac{2450}{3430} = 0.71\,\mathrm{m}$

7 change in gravitational $E_p = mgh$

$96\,000 = m \times 9.8 \times 8$

$m = \dfrac{96\,000}{78.4} = 1224\,\mathrm{kg}$

8 a) $E_K = \dfrac{1}{2}mv^2$

$E_K = \dfrac{1}{2} \times 150 \times 10^6 \times 13^2 = 1.2675 \times 10^{10}$

$E_K = 1.27 \times 10^{10}\,\mathrm{J}$

Note: do not put too many figures in answer

b) Kinetic energy at $16\,\mathrm{m\,s^{-1}}$ is $E_K = \dfrac{1}{2}mv^2$

$E_K = \dfrac{1}{2} \times 150 \times 10^6 \times 16^2$

$E_K = 1.92 \times 10^{10}\,\mathrm{J}$

change in $E_K = E_K(\text{large}) - E_K(\text{small})$

$= (1.92 \times 10^{10}) - (1.27 \times 10^{10})$

$= 0.65 \times 10^{10}\,\mathrm{J} = 6.5 \times 10^9\,\mathrm{J}$

9 $E_K = \dfrac{1}{2}mv^2$

$2.93 \times 10^{10} = \dfrac{1}{2} \times m \times 3100^2$

$m = \dfrac{2 \times 2.93 \times 10^{10}}{9.61 \times 10^6} = 6098\,\mathrm{kg}$

Note: do not put too many figures in answer

10 Note: $260\,\mathrm{kJ} = 260 \times 10^3\,\mathrm{J} = 260\,000\,\mathrm{J}$

$E_K = \dfrac{1}{2}mv^2$

$260 \times 10^3 = \dfrac{1}{2} \times 1000 \times v^2$

$260 \times 10^3 = 500v^2$

$v^2 = \dfrac{260 \times 10^3}{500} = 520$

$v = \sqrt{520} = 22.8\,\mathrm{m\,s^{-1}}$

11 a) Note: $200\,\mathrm{mm} = 200 \times 10^{-3}\,\mathrm{m} = 0.200\,\mathrm{m}$

vertical height $= \dfrac{\text{height of}}{\text{one step}} \times \dfrac{\text{number of}}{\text{steps}}$

vertical height $= 200 \times 10^{-3} \times 30 = 6.0\,\mathrm{m}$

b) gain in gravitational E_p $= mgh = 45 \times 9.8 \times 6 = 2646\,\mathrm{J}$

c) $P = \dfrac{\text{energy transferred}}{\text{time taken}} = \dfrac{\text{gain in } E_p}{t} = \dfrac{2646}{6} = 441\,\mathrm{W}$

d) An underestimate, since Mary will have produced heat and sound during her running up the stairs. (If Mary started from rest then she will also have gained kinetic energy.)

12 a) gain in gravitational $E_p = mgh = 60 \times 9.8 \times 14$
$$= 8232\,J$$

b) i) work done against friction $= F_r \times d = 34 \times 120$
$$= 4080\,J$$

ii) $\dfrac{\text{loss in}}{\text{gravitational } E_p} = \dfrac{\text{gain}}{\text{in } E_K} + \dfrac{\text{work done}}{\text{against friction}}$

$8232 = (E_K(\text{large}) - 0) + 4080$

$E_K(\text{large}) = 8232 - 4080$

$\dfrac{1}{2} \times 60 \times v^2 = 4152$

$v^2 = \dfrac{4152}{30} = 138.4$

$v = \sqrt{138.4} = 11.8\,m\,s^{-1}$

13 a) i) gain in gravitational $E_p = mgh = 50 \times 9.8 \times 3$
$$= 1470\,J$$

ii) $P = \dfrac{E}{t} = \dfrac{\text{gain in } E_p}{t} = \dfrac{1470}{15} = 98\,W$

b) $\dfrac{\text{loss in}}{\text{gravitational } E_p} = \dfrac{\text{gain}}{\text{in } E_K} + \dfrac{\text{work done}}{\text{against friction}}$

$mgh = (E_K(\text{large}) - 0) + (F_r \times d)$

$50 \times 9.8 \times 3 = \left(\dfrac{1}{2} \times 50 \times 1.4^2\right) + (F_r \times 3)$

$1470 = 49 + 3F_r$

$3F_r = 1470 - 49$

$F_r = \dfrac{1421}{3} = 473.667 = 474\,N$

Note: do not put too many figures in answer

14 $\dfrac{\text{loss in}}{\text{gravitational } E_p} = \dfrac{\text{gain}}{\text{in } E_K} + \dfrac{\text{work done}}{\text{against friction}}$

$mgh = (E_K(\text{large}) - E_K(\text{small})) + (F_r \times d)$

$70 \times 9.8 \times 100 = \left(\dfrac{1}{2} \times 70 \times 15^2 - \dfrac{1}{2} \times 70 \times 4^2\right)$
$$+ (F_r \times 500)$$

$68\,600 = (7875 - 560) + 500F_r$

$68\,600 = 7315 + 500F_r$

$500F_r = 68\,600 - 7315$

$F_r = \dfrac{61\,285}{500} = 122.57 = 123\,N$

Note: do not put too many figures in answer

15 a) Note: $40\,mm = 40 \times 10^{-3}\,m = 0.040\,m$

loss in gravitational $E_p = mgh$

loss in gravitational $E_p = 0.3 \times 9.8 \times 40 \times 10^{-3}$

loss in gravitational $E_p = 0.12\,J$

b) gain in E_K of ball = loss in gravitational E_p

$\left(\dfrac{1}{2}mv^2 - 0\right) = 0.12$

Note: pendulum bob is initially at rest, $E_K(\text{small})$
$$= 0$$

$\dfrac{1}{2} \times 0.3 \times v^2 = 0.12$

$v^2 = \dfrac{0.12}{0.15} = 0.8$

$v = \sqrt{0.8} = 0.89\,m\,s^{-1}$

16 a) $E_p = mgh$

$E_p = 150 \times 9.8 \times 25$

$E_p = 36\,750\,J$

b) power $= \dfrac{\text{work done by crane in lifting bucket}}{\text{time taken}}$

power $= \dfrac{\text{gain in } E_p \text{ of bucket}}{\text{time taken}}$

$P = \dfrac{E_p}{t} = \dfrac{mgh}{t} = \dfrac{36\,750}{40} = 918.75 = 919\,W$

Answers to Section 6

12 Space exploration (page 156)

1

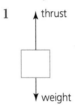

2 a) $W = mg = 1500 \times 9.8 = 14\,700\,N$

Note: g for Earth from *Data Sheet*

b) unbalanced force = thrust − weight
$$= 25\,000 - 14\,700 = 10\,300\,N$$

$F_{un} = ma$

$10\,300 = 1500 \times a$

$a = \dfrac{10\,300}{1500} = 6.9\,m\,s^{-2}$

3 a) unbalanced force = thrust + (− weight)
$$= (3.0 \times 10^7) - mg$$

$F_{un} = (3.0 \times 10^7) - (2.2 \times 10^6 \times 9.8)$
$$= (3.0 \times 10^7) - (2.16 \times 10^7) = 8.4 \times 10^6\,N$$

b) $F_{un} = ma$

$8.4 \times 10^6 = 2.2 \times 10^6 \times a$

$a = \dfrac{8.4 \times 10^6}{2.2 \times 10^6} = 3.8\,m\,s^{-2}$

c) Thrust remains constant but mass of rocket decreases (as fuel is being burnt). Unbalanced force increases as weight of rocket decreases (F_{un} = thrust + (− weight)), so acceleration increases ($a = F_{un}/m$; F_{un} increases and m decreases, both of which make a increase).

Answers

4 a) $W = mg = 4500 \times 1.6 = 7200\,\text{N}$
Note: g for Moon from *Data Sheet*

b) $F_{\text{un}} = \text{thrust} + (-\text{weight}) = (16 \times 10^3) - 7200$
$= 8800\,\text{N}$

$F_{\text{un}} = ma$
$8800 = 4500 \times a$
$a = \dfrac{8800}{4500} = 1.96\,\text{m s}^{-2}$

5 a) Note: $20\,\text{g} = 20 \times 10^{-3}\,\text{kg} = 0.020\,\text{kg}$
$E_K = \dfrac{1}{2}mv^2 = \dfrac{1}{2} \times 20 \times 10^{-3} \times 15\,000^2$
$E_K = 2.25 \times 10^6\,\text{J}$

b) $E_h = cm\Delta T$ Note: c_{iron} from *Data Sheet*
$2.25 \times 10^6 = 480 \times 20 \times 10^{-3} \times \Delta T$
$\Delta T = \dfrac{2.25 \times 10^6}{480 \times 20 \times 10^{-3}} = 234\,375\,°\text{C} = 2.3 \times 10^5\,°\text{C}$

c) The large rise in temperature means that the meteorite would be vaporised.

6 Note: electromagnetic waves travel at $3 \times 10^8\,\text{m s}^{-1}$
Note: $55\,000\,\text{km} = 55\,000 \times 10^3\,\text{m} = 55\,000\,000\,\text{m}$
Note: Journey to satellite $= 2 \times 55\,000 \times 10^3\,\text{m}$ and back
$v = \dfrac{d}{t}$
$3 \times 10^8 = \dfrac{2 \times 55\,000 \times 10^3}{t}$
$t = \dfrac{2 \times 55\,000 \times 10^3}{3 \times 10^8}$
$t = 0.37\,\text{s}$

13 Cosmology (page 164)

1 a) A light year is the distance travelled by light in 1 year.

b) 1 year = 365 days = (365×24) hours
1 year = $(365 \times 24 \times 60)$ minutes
1 year = $(365 \times 24 \times 60 \times 60)$ s
d in 1 year = $(3 \times 10^8) \times (365 \times 24 \times 60 \times 60)$
$= 9.46 \times 10^{15}\,\text{m}$
d in 8.6 years = $8.6 \times 9.46 \times 10^{15} = 8.1 \times 10^{16}\,\text{m}$

2 a) P = microwaves, Q = ultraviolet

b) Gamma rays

3 Coolest Z (red), W (orange–red), Y (yellow) and X (bluish white) hottest

4 The elements that make up the star

5 a) Photographic film or X-ray intensifier

b) Fluorescent materials or photographic film

c) Aerial and microwave receiver

Exam practice for Chapters 9–13 (pages 165–168)

1 a) Susan uses the measuring tape to measure the length of the straight road. When Lynne enters the straight road Susan starts her stopwatch. When Lynne leaves the straight road Susan stops her stopwatch.
$\text{average speed} = \dfrac{\text{length of road}}{\text{time recorded on stopwatch}}$

b) i) $W = mg = 15 \times 9.8 = 147\,\text{N}$
The minimum force needed to lift the bicycle at constant speed is the force that balances the weight of the bicycle.
Minimum force required to lift = $147\,\text{N}$

ii) Note: 10 steps = $10 \times 80\,\text{mm} = (10 \times 80 \times 10^{-3})\,\text{m}$
$E_W = F \times d = 147 \times (10 \times 80 \times 10^{-3}) = 118\,\text{J}$

2 a) $a = \dfrac{v-u}{t} = \dfrac{0-40}{28} = \dfrac{-40}{28} = -1.4\,\text{m s}^{-2}$

b) distance = area under speed–time graph
distance = area of rectangle + area of triangle
$\text{distance} = (1 \times 40) + \left(\dfrac{1}{2} \times 28 \times 40\right)$
$= 40 + 560 = 600\,\text{m}$
Yes, the train stops within this distance. It takes 600 m to stop. This is 20 m short of the signal.

3 a) 1200 N. Since the ship is moving at constant speed (Newton's first law), the forces acting on it must be balanced.

b) i) $F_{\text{un}} = ma$
$-1200 = 240\,000 \times a$
$a = \dfrac{-1200}{240\,000} = -0.005\,\text{m s}^{-2}$

ii) $a = \dfrac{v-u}{t}$
$-0.005 = \dfrac{0-1.5}{t}$
$-0.005 = \dfrac{-1.5}{t}$
$t = \dfrac{-1.5}{-0.005} = 300\,\text{s}$

iii) $\text{average velocity} = \dfrac{u+v}{2} = \dfrac{0+1.5}{2} = 0.75\,\text{m s}^{-1}$
Note: this equation may be used because acceleration is constant.
$\text{average velocity} = \dfrac{\text{displacement}}{\text{time taken}}$
$0.75 = \dfrac{\text{displacement}}{300}$

displacement $= 0.75 \times 300 = 225\,\text{m}$

or

distance $=$ area under speed–time graph

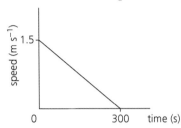

distance $= \dfrac{1}{2} \times 300 \times 1.5 = 225\,\text{m}$

4 a) gain in gravitational $E_P = mgh$
$$= 25 \times 9.8 \times (5.5 - 0.5)$$
$$= 1225\,\text{J}$$

Note: $h = (5.5 - 0.5)$

  b) i) Note: 5 minutes $= (5 \times 60)\,\text{s}$
      gain in E_P for 20 bales $= 1225 \times 20 = 24\,500\,\text{J}$

      power $P = \dfrac{\text{energy transferred}}{\text{time taken}} = \dfrac{\text{gain in } E_P}{t}$

      $= \dfrac{24\,500}{(5 \times 60)} = 81.7\,\text{W}$

    ii) Energy will be transferred to the surroundings, in the form of heat and perhaps sound, during the lifting process. (The bales will also gain some kinetic energy when put onto the elevator.)

5 a) gain in gravitational $E_P = mgh$
$$= 0.2 \times 9.8 \times 0.3 = 0.588\,\text{J}$$

  b) i) Gravitational potential energy is changed into kinetic energy and some heat (due to work done against the resistive force).

    ii) Assuming that no energy is transferred to the surroundings:

      gain in $E_K =$ loss in gravitational E_P

      $\dfrac{1}{2}mv^2 - 0 = mgh$ Note: initial $E_K = 0 = E_K(\text{small})$

      $\dfrac{1}{2}mv^2 = 0.588$

      $\dfrac{1}{2} \times 0.2 \times v^2 = 0.588$

      $v^2 = \dfrac{0.588}{0.1} = 5.88$

      $v = \sqrt{5.88} = 2.42\,\text{m s}^{-1}$

6 a) change in gravitational $E_P = mgh$
$$= 45 \times 9.8 \times 6 = 2646\,\text{J}$$

  b) gain in $E_K =$ loss in gravitational E_P

    $\dfrac{1}{2}mv^2 - 0 = mgh$ Note: initial $E_K = 0 = E_K(\text{small})$

    $\dfrac{1}{2}mv^2 = 2646$

    $\dfrac{1}{2} \times 45 \times v^2 = 2646$

    $v^2 = \dfrac{2646}{22.5} = 117.6$

    $v = \sqrt{117.6} = 10.8\,\text{m s}^{-1}$

  c) work done against friction $=$ loss in E_K

    $F_r \times d = 2646\,\text{J}$

    $15 \times d = 2646$

    $d = \dfrac{2646}{15} = 176\,\text{m}$

7 a) i) $a = \dfrac{v - u}{t} = \dfrac{4.5 - 0}{30} = \dfrac{4.5}{30} = 0.15\,\text{m s}^{-2}$

    ii) distance $=$ area under velocity–time graph

      $d = \left(\dfrac{1}{2} \times 30 \times 4.5\right) + (120 \times 4.5) + (30 \times 3) +$

      $\left(\dfrac{1}{2} \times 30 \times 1.5\right)$

      $d = 67.5 + 540 + 90 + 22.5 = 720\,\text{m}$

    iii) average speed $= \dfrac{d}{t} = \dfrac{720}{180} = 4.0\,\text{m s}^{-1}$

  b) i) Unbalanced, since boat is accelerating (Newton's second law)

    ii) Balanced, since the boat is travelling at constant velocity (Newton's first law)

8 a) i) Constant acceleration from rest to $4.9\,\text{m s}^{-1}$ in $0.5\,\text{s}$

    ii) Constant acceleration from $-2.9\,\text{m s}^{-1}$ to rest in $0.3\,\text{s}$

  b) displacement $=$ area under velocity–time graph from $0\,\text{s}$ to $0.5\,\text{s}$

    displacement $=$ area of triangle $= \dfrac{1}{2} \times 0.5 \times 4.9$

    displacement $= 1.23\,\text{m}$

  c) displacement $=$ area under velocity–time graph from $0.5\,\text{s}$ to $0.8\,\text{s}$

    displacement $=$ area of triangle $= \dfrac{1}{2} \times 0.3 \times -2.9$

    displacement $= -0.44\,\text{m}$

    height $= 0.44\,\text{m}$

  d) During the collision with the ground some of the kinetic energy of the ball is changed into heat and sound. The ball rebounds with less kinetic energy (slower speed) and so rises to a lower height.

  e) $0.5\,\text{s}$, since the velocity changes sign showing that the ball is moving in the opposite direction.

9 a) Using a vector diagram:

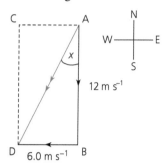

resultant velocity = AD

$AD^2 = AB^2 + BD^2 = 12^2 + 6^2 = 144 + 36 = 180$

$AD = \sqrt{180} = 13.4\,\mathrm{m\,s^{-1}}$

$\tan x = \dfrac{BD}{AB} = \dfrac{6}{12} = 0.5$

$x = 27°$

resultant velocity of boat = $13.4\,\mathrm{m\,s^{-1}}$ at 27° W of S

If using a scale vector diagram then

$13.4 \pm 0.5\,\mathrm{m\,s^{-1}}$ at $27 \pm 2°$ W of S

b) 1200 N as there must be balanced forces acting on the boat as it is travelling at constant speed.

10 a) horizontal speed = $\dfrac{\text{horizontal distance}}{\text{time taken}}$

Note: horizontal motion is a constant speed of $2.5\,\mathrm{m\,s^{-1}}$

$2.5 = \dfrac{\text{horizontal distance}}{3.2}$

horizontal distance = $2.5 \times 3.2 = 8.0\,\mathrm{m}$

b) $a = \dfrac{v - u}{t}$ Note: vertical motion is a constant acceleration of $9.8\,\mathrm{m\,s^{-2}}$ from rest

$9.8 = \dfrac{v - 0}{3.2}$

$v = 9.8 \times 3.2 = 31.4\,\mathrm{m\,s^{-1}}$

c) average vertical speed = $\dfrac{u + v}{2}$ Note: this equation may be used because acceleration is constant

average vertical speed = $\dfrac{0 + 31.4}{2} = 15.7\,\mathrm{m\,s^{-1}}$

height = vertical distance travelled

height = average vertical speed × time taken

height = $15.7 \times 3.2 = 50.2\,\mathrm{m}$

or

height = area under vertical velocity–time graph

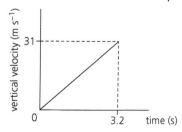

height = $\dfrac{1}{2} \times 3.2 \times 31.4 = 50.2\,\mathrm{m}$

11 a) The tennis ball has two independent motions taking place at the same time:
- a horizontal motion with a constant velocity of $25\,\mathrm{m\,s^{-1}}$
- a vertical motion with a constant acceleration (downwards) of $9.8\,\mathrm{m\,s^{-2}}$ starting from rest.

The resultant velocity of the tennis ball continually changes in magnitude and direction and the path followed is curved.

b) horizontal distance travelled = area under horizontal velocity–time graph

horizontal distance travelled = $0.7 \times 25 = 17.5\,\mathrm{m}$

c) height = vertical distance travelled

height = area under vertical velocity–time graph

vertical distance travelled = $\dfrac{1}{2} \times 0.7 \times 6.9 = 2.42\,\mathrm{m}$

12 a) Altitude of communications satellite is greater than that of the polar satellite.

b) i) A satellite that has a period of 24 hours.

ii) Note: electromagnetic waves travel at $3 \times 10^8\,\mathrm{m\,s^{-1}}$

Note: $36\,000\,\mathrm{km} = 36\,000 \times 10^3\,\mathrm{m} = 36\,000\,000\,\mathrm{m}$

Note: Journey to satellite = $2 \times 36\,000 \times 10^3\,\mathrm{m}$ and back

$v = \dfrac{d}{t}$

$3 \times 10^8 = \dfrac{2 \times 36\,000 \times 10^3}{t}$

$t = \dfrac{2 \times 36\,000 \times 10^3}{3 \times 10^8}$

$t = 0.24\,\mathrm{s}$

13 a) Thrust of the engine (upwards) is bigger than the weight (downwards). This means that there is an unbalanced force (upwards) which accelerates the rocket (upwards).

b) $W = mg = (2.1 \times 10^6 + 2.3 \times 10^3) \times 9.8 = 2.06 \times 10^7\,\mathrm{N}$

$F_{un} = \text{thrust} + (-\text{weight})$

$= (2.5 \times 10^7) - (2.06 \times 10^7) = 4.4 \times 10^6\,\mathrm{N}$

$F_{un} = ma$

$4.4 \times 10^6 = [(2.1 \times 10^6) + (2.3 \times 10^3)] \times a$

$a = \dfrac{4.4 \times 10^6}{2.1 \times 10^6} = 2.1\,\mathrm{m\,s^{-2}}$

c) $E_K = \dfrac{1}{2}mv^2 = \dfrac{1}{2} \times 2.3 \times 10^3 \times (1200)^2 = 1.66 \times 10^9\,\mathrm{J}$

d) Note: $900\,km = 900 \times 10^3\,m = 900\,000\,m$

$$v = \frac{d}{t}$$

$$3 \times 10^8 = \frac{900 \times 10^3}{t} \quad \text{Note: EM waves travel at}$$
$$3 \times 10^8\,ms^{-1}$$

$$t = \frac{900 \times 10^3}{3 \times 10^8} = 0.003\,s$$

e)
$$v = f\lambda$$
$$3 \times 10^8 = 1.06 \times 10^{15} \times \lambda$$
$$\lambda = \frac{3 \times 10^8}{1.06 \times 10^{15}} = 2.8 \times 10^{-7}\,m$$

f) Note: $60\,mA = 60 \times 10^{-3}\,A = 0.060\,A$
$$P = IV = 60 \times 10^{-3} \times 0.8 = 0.048\,W$$

14 a) change in $E_K = E_K(\text{large}) - E_K(\text{small})$

$$= \left(\frac{1}{2} \times 750 \times (380)^2\right) -$$

$$\left(\frac{1}{2} \times 750 \times (170)^2\right)$$

change in $E_K = 4.33 \times 10^7\,J$

b) $E_h = cm\Delta T$

$$4.33 \times 10^7 = 1400 \times 750 \times \Delta T$$

$$\Delta T = \frac{4.33 \times 10^7}{1.05 \times 10^6} = 41\,°C$$

c) i) energy supplied $= P \times t = 2400 \times 130$
$$= 312\,000\,J$$

Note: $600\,g = 600 \times 10^{-3}\,kg = 0.600\,kg$

energy absorbed $= ml$

$$312\,000 = 600 \times 10^{-3} \times l$$

$$l = \frac{312\,000}{600 \times 10^{-3}} = 5.2 \times 10^5\,J\,kg^{-1}$$

ii) Some of the energy supplied by the heater will be transferred to the surroundings. The sample will therefore absorb less than $312\,000\,J$ of energy. Hence the value calculated for the specific latent heat of fusion of the sample will be too high.